The Palgrave Lacan Series

Jacques Lacan is one of the most important and influential thinkers of the 20th century. The reach of this influence continues to grow as we settle into the 21st century, the resonance of Lacan's thought arguably only beginning now to be properly felt, both in terms of its application to clinical matters and in its application to a range of human activities and interests. The Palgrave Lacan Series is a book series for the best new writing in the Lacanian field, giving voice to the leading writers of a new generation of Lacanian thought. The series will comprise original monographs and thematic, multi-authored collections. The books in the series will explore aspects of Lacan's theory from new perspectives and with original insights. There will be books focused on particular areas of or issues in clinical work. There will be books focused on applying Lacanian theory to areas and issues beyond the clinic, to matters of society, politics, the arts and culture. Each book, whatever its particular concern, will work to expand our understanding of Lacan's theory and its value in the 21st century.

More information about this series at
http://www.palgrave.com/gp/series/15116

Clint Burnham • Paul Kingsbury
Editors

Lacan and the Environment

palgrave
macmillan

Editors
Clint Burnham
English Department
Simon Fraser University
Vancouver, BC, Canada

Paul Kingsbury
Faculty of Environment
Simon Fraser University
Burnaby, BC, Canada

The Palgrave Lacan Series
ISBN 978-3-030-67204-1 ISBN 978-3-030-67205-8 (eBook)
https://doi.org/10.1007/978-3-030-67205-8

Cover illustration: Westend61/gettyimages

This Palgrave Macmillan imprint is published by the registered company Springer Nature Switzerland AG.
The registered company address is: Gewerbestrasse 11, 6330 Cham, Switzerland

Acknowledgments

This book is largely the result of LaConference 2018—"Lacan and the Environment"—which was organized by the Lacan Salon and held at the Western Front in Vancouver, British Columbia, on the unceded, traditional territories of the Coast Salish Peoples, including the Squamish (Sḵwx̱wú7mesh Úxwumixw), Tsleil-Waututh (səl̓ilw̓ətaʔɬ) and Musqueam (xʷməθkʷəy̓əm) Nations. We are extremely grateful to the Department of English, the Department of Geography, Faculty of Environment, the Faculty of Arts and Social Sciences, the Faculty of Communication, Art and Technology at Simon Fraser University. Thank you as well to Derek Hook and Calum Neill for their support for this project, to Beth Farrow and Joanna O'Neil at Palgrave for shepherding the manuscript, to Sam Brawand for editorial assistance, and to Alois Sieben for the indexing. Thank you, finally, to our colleagues in the Lacan Salon, which has been an unsettled and jouissance-ridden space for psychoanalytic thinking in Vancouver since 2007.

Praise for *Lacan and the Environment*

"*Lacan and the Environment* is today's version of what Hegel called "infinite judgment," the assertion of the link between two dimensions which appear totally different. This outstanding volume throws a new light not only on Lacan but also on environmental issues: we cannot really understand ecology without taking into account all the fantasies that overdetermine our approach to this topic."

—Slavoj Žižek, *Department of Philosophy, University of Ljubljana and International Director of the Birkbeck Institute for the Humanities of the University of London, UK*

"The defining crisis of our time is ecocide. Upending conventional distinctions between exterior and interior, psychoanalysis proves an unexpectedly rich resource for thinking and acting against destruction, displacement, and extinction. These smart, urgent essays consider a broad range of cultural contexts, illustrate the centrality of fantasy, desire, and symbolization to ecological transformation, and should inspire and terrify readers of many stripes."

—Anna Kornbluh, *Department of English, University of Illinois, Chicago, USA*

"This brilliant edited volume not only reveals the environment to be an enduring theme in Lacan's oeuvre, but also rethinks and reworks Lacan environmentally, showing "nature" to be a site of both play and anxiety, interiority and radical externality, pleasure and pollution. Our study of the environment will never be the same."

—Ilan Kapoor, *Faculty of Urban and Environmental Change, York University, Toronto, Canada*

Contents

Notes on Contributors

Clint Burnham is Professor of English and chair of the graduate program (English) at Simon Fraser University, Vancouver, Canada, and president of the Lacan Salon. His recent books include *Does the Internet have an Unconscious? Slavoj Žižek and Digital Culture* (Bloomsbury, 2018) and *Fredric Jameson and* The Wolf of Wall Street (Bloomsbury, 2016).

Alessandra Capperdoni is lecturer in the Department of Humanities at Simon Fraser University, Canada, where she teaches literature, philosophy and critical thought, and animal studies. Her research explores desire, affect, trauma, and social imaginaries in three interrelated areas: the processual forms of language-oriented poetics and translation practices through the lens of the theories of the unconscious of Freud and Lacan, Marxist-inflected philosophies of language, poststructuralism and feminist theory; the work of literature and art in relation to violence and trauma; cultural work as phenomenological orientation toward the environment and other-than-human animals, engendering the possibility of social links with different worlds.

Juan Luis de la Mora practices psychoanalysis in Mexico City. His interests focus mainly on psychoanalysis as a praxis of counter-hegemonic subjectivation, as well as subjectivities in Western modernity and its clinical manifestations such as anxiety and addictions in light of contemporary capitalism. He has a Bachelor's Degree in Clinical Psychology from

Universidad Iberoamericana and a Master's Degree in Critical Theory from 17, Instituto de Estudios Críticos. In 2016 he published To think of nothing: From consumption to addiction (Casa Editorial Abismos), a book about contemporary subjectivity, addiction, anxiety and capitalism.

Hilda Fernandez-Alvarez works as a Lacanian psychoanalyst in private practice in Vancouver. She has vast clinical experience and her academic background is in clinical psychology and literature. She is a PhD candidate in the Department of Geography at Simon Fraser University (SFU), with a research study on discursive spaces of trauma and healing within the mental health institution. She co-founded the Lacan Salon in 2007 and currently serves as its clinical director. She is an academic associate with the Institute for the Humanities at SFU and leads a Clinical Seminar in Vancouver since the fall 2015. She has published various articles on psychotherapy and psychoanalysis.

Matthew Flisfeder is Associate Professor of Rhetoric and Communications at The University of Winnipeg, Canada. He is the author of *Algorithmic Desire: Toward a New Structuralist Theory of Social Media* (2021), *Postmodern Theory and Blade Runner* (2017), *The Symbolic, The Sublime, and Slavoj Žižek's Theory of Film* (2012), and co-editor of *Žižek and Media Studies: A Reader* (2014).

Nathan Gorelick is Associate Professor of English and Literature at Utah Valley University. He has published numerous articles on psychoanalysis both critical and clinical, in journals including *Continental Thought & Theory*, *Umbr(a): A Journal of the Unconscious*, *Discourse*, *CR: The New Centennial Review*, *Theory & Event*, and *International Journal of Counseling Psychology*. His work also appears in various essay collections, most recently *Subject Lessons: Hegel, Lacan, and the Future of Materialism* (Northwestern UP, 2020) and *Castration, Impotence, and Emasculation in the Long Eighteenth Century* (Routledge, 2020). He is the editor of Provocations Journal, affiliated with the University of Nebraska Press book series of the same name.

Paul Kingsbury is Professor of Geography and Associate Dean of the Faculty of Environment at Simon Fraser University, and Vice President of the Lacan Salon. His most recent scholarship draws on psychoanalysis in order to investigate the cultural geographies of paranormal investigations. He is the co-editor (with Anna J. Secor) of *A Place More Void* (2021) and (with Steve Pile) *Psychoanalytic Geographies* (2014).

Alma Krilic holds a PhD in Philosophy of Education and teaches at Simon Fraser University in the Continuing Studies Department. Her interests include Lacanian psychoanalysis, language education and philosophy of education. Alma has been a member of Lacan Salon since 2013 and currently serves as Lacan Salon Secretary.

Sasha J. Langford is a writer, composer, and interdisciplinary artist residing as a settler in Vancouver, B.C., the occupied territories of the xʷməθkʷəy̓əm (Musqueam), Sḵwx̱wú7mesh (Squamish), and səl̓ilwətaʔɬ (Tsleil-Waututh) peoples. With her sound-based performance work, Sasha has exhibited across Canada, the United States, and New Zealand, and is the organizer of Social/Movements, a reading group centered on dance writing and theory. She holds an M.A. in Media Studies from Concordia University, where her research considered landscape photography from a psychoanalytic and anti-colonial perspective. She teaches media history and theory at Columbia College in Vancouver.

Calum Matheson is Assistant Professor of Public Deliberation and Civic Life in the Department of Communication at the University of Pittsburgh, director of the William Pitt Debating Union, and a candidate at the Pittsburgh Psychoanalytic Center. His work focuses on intersections of Lacanian psychoanalysis, cultural studies, and rhetoric. He is author of *Desiring the Bomb: Communication, Psychoanalysis, and the Atomic Age*. His articles have been published in *Quarterly Journal of Speech, Review of Communication, Rhetoric Society Quarterly, Philosophy and Rhetoric*, and other leading journals.

Todd McGowan teaches theory and film at the University of Vermont. He is the author of *Universality and Identity Politics, Emancipation After Hegel, Only a Joke Can Save Us: A Theory of Comedy, Capitalism and Desire*, and other works. He is the coeditor of the Diaeresis series along with Slavoj Žižek and Adrian Johnston at Northwestern University Press and the editor of the Film Theory in Practice series at Bloomsbury.

Tamás Nagypál is a lecturer at Ryerson University's School of Image Arts. He holds a PhD in Cinema and Media Studies from York University, and he is currently working on book manuscript titled *The Dark Passage to Human Capital: Film Noir and Neoliberalism*. His publications include articles in the journals *Film International*, *The Journal of Religion and Film*, and *Mediations*, as well as book chapters in the edited volumes *Zizek and Media Studies: A Reader* and *Monstrous Children and Childish Monsters: Essays on Cinema's Holy Terrors*.

Miguel Rivera is a PhD candidate at Tuft University's Department of English. His research interests include African-American literature, genre fiction, film studies, and psychoanalysis, and literary theory more broadly. His recent publications include "'[D]ifferent even from our 'own' differences': Racial Signification and the Legacy of Lacanian Sexuation" in *Word and Text: A Journal of Literary Studies and Linguistics* and "'A Critique of Countertransference': Commentary on Session XII" in *Reading Lacan's Seminar VIII: Transference*.

Alois Sieben is a PhD candidate in the English Department at Simon Fraser University. His SSHRC-funded dissertation uses psychoanalytic theory to examine representations of digital search technology in film, fiction and poetry, arguing that the Internet orientates itself around a constant excavation of the "unsearchable." He has published work recently in *Cineaction*, as well as appeared on Vanessa Sinclair's *Rendering Unconscious* podcast.

Cindy Zeiher is a lecturer at the University of Canterbury, New Zealand. Her writings and publications are in the areas of Lacanian psychoanalysis, politics, subjectivities and critical theory. Together with Todd McGowan she has edited and contributed to a collection of essays, *Can Philosophy Love? Reflections and Encounters* (2017, Rowman and Littlefield International) and has co-authored a book with Ed Pluth, *On Silence: Holding the Voice Hostage* (2019, Lacan Palgrave Series). She is currently editing a collection of essays, *Psychoanalysis and Stupidity* (Rowman and Littlefield International, forthcoming 2021) and is co-editor along with Mike Grimshaw, of the journal, *CT&T: Continental Thought and Theory*.

List of Figures

1

Introduction

Clint Burnham and Paul Kingsbury

We begin this book with a psychoanalytic provocation: do we think of the environment as our mother, or as a child? That is, to be Oedipal, do we want to make love to the earth, or do we want to protect it? Or perhaps the problem is that in both cases we want to protect the earth, a fantasy that fails to take into account our libidinal investment. The mother was here before us, we have to take care of her as she dies (this is the argument around the new master signifier, the Anthropocene), or it is our future, we have to save it (rising sea levels mean our child is drowning)? If the earth is our mother—and this fantasy ranges from spurious New Age notions of Gaia to more political concepts in Indigenous knowledges—there are different kinds of mothers. Mothers protect us

C. Burnham (✉)
English Department, Simon Fraser University, Vancouver, BC, Canada
e-mail: clint_burnham@sfu.ca

P. Kingsbury
Faculty of Environment, Simon Fraser University, Burnaby, BC, Canada
e-mail: kingsbury@sfu.ca

C. Burnham, P. Kingsbury (eds.), *Lacan and the Environment*, The Palgrave Lacan Series, https://doi.org/10.1007/978-3-030-67205-8_1

but they are also a threat. They smother us—this is the mother that Jacques Lacan tells us from whom we have to extricate ourselves, the maternal crocodile or she-wolf or praying mantis. And these figures, in Lacan, drawn as they are from the animal world, are neither patently misogynist nor unproblematic turns to nature. Rather, they indicate a fundamental antagonism or contradiction in our attitudes to the person who gave birth to us and nurtured us. We resent the earth, because we depend on it.

And the earth, it turns out, depends on us. Again, our attitude is ambivalent. To grapple with this ambivalence, we can work through some of the logic of the earth as child via Sigmund Freud's analysis of the fantasy "a child is being beaten."[1] Remarking how often his patients "confess to having indulged in the phantasy: 'A child is being beaten,'" Freud lists a variety of permutations: from a boy being beaten by the father, to a boy being beaten by the mother, to a girl being beaten by the father. Moreover, Freud wants us to attend to disavowed enjoyment. What does this mean for the environmentalist subject? What are the unconscious anxieties and fantasies that subvent contemporary environmentalism? How can such a politics be confronted in terms of its enjoyment, its *jouissance*, in order to understand better why, and how, we oppose resource extraction?

That is, thinking of the environment as mother or child—or, rather, pointing out that our conflicted cathexes in family dynamics underly our attitudes towards nature and the climate crisis—suggests a certain psychoanalytic universality. We all, presumably, have mothers, and are or were children. But psychoanalysis is also the science of the particular, and we each have specific ways in which we act with respect to the environmental crisis; we each have specific ways in which we relate to others in this particular moment and locales. We act in different ways: we recycle, or we prep for an apocalypse, we compulsively check that our pet has enough water, or we save for a Tesla (or a Hummer).[2] And we congratulate ourselves on our recycling, our environmental "beautiful soul," as Hegel put it. We are pragmatic, like doomsday preppers (as Calum Matheson describes in his chapter), or we even enjoy our large vehicles and consumerism. But much of this activity is surely a matter of keeping busy, of obsessively doing nothing—and less such a diagnosis seem beside the point, it is worth keeping in mind the verbal attacks on climate activist Greta Thunberg for her Asperger's syndrome, which she refers to as her "superpower" for how it helps her attend to her politics.[3]

The "beautiful soul" adheres to the masculine logic of the exception—I am a good person because I recycle. This logic, as worked out in Lacan's formula of sexuation in *Seminar XX*, depends for its self-congratulation on others who do *not* recycle. The social link is thus fraught. A different logic, or subjectivity, is at work with climate activisms and protests. Or, rather, while the "beautiful soul" certainly is to be found, the protestor is more precisely a hysterical subject, appealing to the master: government, corporations.[4] Environmentalism is a social activity: which means that for Lacan the problem is others, both the big Other, or authority figures, and the small other, or other people.

How we think about the environment, about nature, in the age of climate crisis and the Anthropocene, has psychoanalytic resonances: we take an unconscious enjoyment in *not* acting, or we wonder why others (climate deniers, corporations, and governments) do not listen to science, do not act. Lacan can help us work through these conflicted desires and anxieties. Consider a basic tripartite division in Lacanian theory: the Imaginary, the Symbolic, and the Real. The Imaginary environment is the space around us, as Lacan remarks in the "mirror stage" essay, describing how the child "playfully experiences the relationship between the movements made in the image and the reflected environment."[5] That is, here the environment is an image, it is constitutive of the ego and of our narcissistic relation with others. The environment is a site of play, of a "flutter of jubilant activity" as Lacan later says, but also of alienation and aggressivity.[6] These dualisms of affect are crucial if we are to understand the differing ways in which we relate to and enjoy the environment, and if we are not to descend into moralistic dismissals of those with whom we disagree. The environment, or nature, in terms of the imaginary is also nature as beauty, breathtaking, even sublime.

In terms of the Symbolic, which is to say order, the Law, and language, the environment is that as described and analysed by science: categorisation, classification, measurement, and exploration. But present-day scientific claims, or what Lacan refers to as the University discourse, suffer from a credibility problem. As the Wikipedia page for Slavoj Žižek points out, today's "postmodern subject is cynical toward official institutions, yet at the same time believes in conspiracies."[7] That is to say, there is no big Other, science cannot be trusted, and yet we are all gullible, falling for misleading images of Amazon forest fires.[8] Science without the big Other, without the Law, is a discourse of proliferating regulations big and small, from recycling directives to anxieties about which limit of global warming we can avert: 2.5 degrees (F or C) by 2100? Governmentalism does not mean popular decision-making, but rather global elites meeting in attractive cities like Madrid, Copenhagen, or Paris. Climate conferences rarely take place in dirty places like Newark, Lagos, or Leeds, for pollution may be too close to the Real, or that which resists language, conceptualisations, or understanding.

Marc De Kesel can help us think about the environment as the Real. He writes that with *Das Ding*, Lacan

> names desire's inability to really come full circle and gives it a specific locus in the libidinal economy. As a radical exteriority, the "thing" nevertheless forms the central point around which the whole economy revolves and at which desire aims, a "topological" paradox for which he forges the term "extimité." While the "thing" indicates the point at which desire aims, as if at "itself," it is also what would destroy it as soon as it was reached. This is why the "thing" is an indispensable element in the cartography of desire.[9]

The "thing," or *Das Ding*, is also, we argue, crucial to understanding the Environment as the Real. For nature as the "thing" is nature that can only be thought of topologically, not outside us, not inside us, but both at the same time. Nature is "extimate": on the one hand, there is a foreign kernel in our innermost being, our unconscious, which we sometimes think of as unnatural; on the other hand, nature functions as a radical externality, the locus of our most sincere feelings and beliefs. Nature, or the environment, is not "out there"—it is how we *relate* to what is "out there"—and to what is "in here." For our relation to nature is what the late Lacan calls a non-relation, one best understood via topological models as argued in Sasha Langford's chapter.

Lacan first developed his concept of the Real in *Seminar XI.* In a well-known passage, he describes a vacation he took in the 1920s:

> Being a young intellectual, I wanted desperately to get away, see something different, throw myself into something practical, something physical, in the country say, or at the sea. One day, I was on a small boat, with a few people from a family of fishermen in a small port. At that time, Brittany was not industrialized as it is now. There were no trawlers. The fisherman went out in his frail craft at his own risk. It was this risk, this danger, that I loved to share. But it wasn't all danger and excitement—there were also fine days. One day, then, as we were waiting for the moment to pull in the nets, an individual known as Petit-Jean, that's what we called him—like all his family, he died very young from tuberculosis, which at that time was a constant threat to the whole of that social class—this Petit-Jean pointed out to me something floating on the surface of the waves. It was a small can, a sardine can. It floated there in the sun, a witness to the canning industry, which we, in fact, were supposed to supply. It glittered in the sun. And Petit-Jean said to me—*You see that can? Do you see it? Well, it doesn't see you!*[10]

The sardine can is garbage: it denotes a loop from the fish (or nature), and their extraction and processing into a canned good, which is then discarded (the packaging—perhaps it fell off the side of a truck or boat). Does garbage see us? Petit-Jean thinks not, and Lacan soon concludes that he himself, as an intellectual intruding on this dangerous manual labour, is out of place in the picture. Perhaps the sardine can is also out of place—or is it? It calmly floats in the water, glitters off the sunlight, and yet it is waste—industrial waste. Metonymically, it connects to today's concerns with the plastic gyre in the North Pacific—a patch of microplastic waste the size of Texas, we are told. But the visual matters as much as place: we or Lacan could see the sardine can, and yet now we cannot see the microplastics or nanoplastics that permeate even sea salt and fish guts—we are talking about anxieties of scale and the gaze.

It is also important to note that Lacan gestures to a history of industrialisation of the fisheries: "At that time, Brittany was not industrialized as it is now. There were no trawlers"—the oceans presumably could not be over-fished as they have been in the past few decades.[11] And this is connected directly to the workers' danger. The logic is this: first, we over-fish the oceans, then, or as part of that process, we despoil them with plastic, fishing nets, and other detritus. In some ways, our attitudes towards workers and nature are similar: we care about them only because they are under threat, because they are suffering. But the fishermen's life-threatening work on the ocean also points to a fundamental human struggle with nature. That struggle, or nature as simultaneously a place of work, then contrasts with Lacan's avowed vacation plans. We cannot hope to fully understand nature, or the environment, unless we understand this contradiction—that nature is a site for labour, and for pleasure, for production, and for consumption.

We can also read the sardine can as an indication of a certain lack in pollution. The sardine can is trash, is litter. In his seminar on Poe's "Purloined Letter," Lacan referred to Joyce's phrase "a letter, a litter" and goes on to describe police, in Poe's story, ignoring a "seeming scrap of waste paper *[dechet]*."[12] Litter, waste, and trash all have different connotations: the first is what we have unthinkingly discarded; the second denotes something extraneous, a by-product; while the third suggests the lowest place in a hierarchy or status: defilement, filth. That is, to see the sardine can as self-evidently garbage, as without lack, is to think of pollution as without lack, as wholly present substance, rather than as split subject. Pollution lacks, it is castrated: the litter is also a letter, the letter of the law, a language we obey. What we mean by this is that pollution is too much like nature: it is indifferent to the human world and our well-intentioned laws and regulations. Acid rain, the radioactive cloud from Chernobyl or waste from Fukushima, ignore national boundaries or international conventions. Lacan's fable of the sardine can help us to think about the environment, and also warns us to be careful in our extracting from his writings.[13]

Nonetheless, what readers will find in this collection is how much themes of the environment, and nature, and indeed garbage, permeate Lacan's work. In *Seminar I*, Lacan affirms the pedagogical value of geology, for instance. "My dear fellows," he remarks, "you wouldn't believe what you owe to geology. If it weren't for geology, how could one end up thinking that one could move, on the same level, from a recent to a much more ancient layer?"[14] Geology, he tells us, is a field of study that illuminates the coexistence of different times in one place. A strength of this collection is that the contributors in their critical endeavours do not simply treat psychoanalysis as a method for deciphering our conflicted affects regarding the environment. For here, environment refers to those affective relationships themselves, to networked existences, unequal exchanges, and implacable conflicts. Our contributors' chapters also argue that environment is not merely understood as nature, and concerns about the environment do not automatically mean the crisis of climate change. We do not extract from Lacan, but instead the book reworks Lacanian concepts, indeed, such questions as how to read Lacan, and how to interpret his ideas through the lens of the environment. Perhaps the sardine can *did* see Lacan, after all!

The collection begins, in Part I, with some fundamental questions of Lacanian theory and how they help us to think about the environment. In Cindy Zeiher's "*Love Thy Enemy:* Environment(al) Politics," she asks *what does the environment want?*, via such key psychoanalytic notions as the neighbour, extimacy, and interpassivity, as well as more properly political concepts, such as sovereignty. As Zeiher persuasively argues, if the environment qua subject is indifferent to us, how do we negotiate our enjoyment in contrast to that of our neighbour, the climate change denier? Zeiher introduces a key distinction—but also a blurring of that distinction, when she posits that politics itself constitutes an environment—and thus a Schmittian friend/enemy both helps us navigate our desire with respect to nature, or the planet earth, even as the extimate proximity of their desire threatens our own.

This subject-other relationship we also visit in Alois Sieben's chapter. Sieben theorizes the Inuit multimedia artist Tanya Tagaq's writing and throat singing in terms of the "environmental *objet petit a*," which involves the collapsing of the boundary between the human and the natural environment. Specifically, Tagaq's throat singing, which shifts between traditional and avant-garde sonic registers, incites "epistemological anxiety," especially for settler audiences. Sieben defines such an anxiety as the "ecological foundation" of the *objet petit a*, which underpins a topological relationship between the subject and the Other. In making his argument, Sieben provides an ecological reading of *Seminar X*, wherein Lacan attends to the "Other environment," and also puts into conversation Lacan and Bruno Latour to consider how the *objet petit a* provides a way of theorizing disconnects between affect and the environmental crisis.

Ultimately for Sieben, the only tenable way of dealing with this crisis is by embracing what Lacan called the "grounded certainty" of anxiety.

Alma Krilic's chapter focuses on the work environment in Herman Melville's short story, "Bartleby, the Scrivener." Krilic considers how Bartleby enjoys his environment in terms of the inactivity of "preferring not" to work. Rather than affirm or denigrate such a refusal as a type of passive resistance, Krilic takes seriously the interior environment of Bartleby's office, which crucially involved the comings and goings of other people. Krilic asks, why did Bartley remain in this spatially confined environment given his lack of pleasure and purpose. Extending Alenka Zupančič engagement with Bartleby's politics of non-preference, the answer for Krilic concerns Bartleby's fundamental enjoyment of the environment in terms of not simply the office space but also the indeterminacy of his addressee. Such a symptom, which Krilic suggests could be masochistic, takes place in a "third space," that is, a "no-man's land neither adapting nor resisting his environment," suggesting a new model for ecological awareness.

Bartleby is famous for preferring "not to," and Clint Burnham's chapter on Lacan's "trash talk" pays similar attention to utterances and language, finding a parallel between online garbage and how we sort through the Seminars for the "gold" of Lacanian *aperçus.* Internet garbage, Burnham argues, can be thought of in terms of the three objects: the phallic object (Φ), the barred Other S(Ⱥ), and the *objet petit a.* He takes as his cultural object a documentary film, *The Cleaners*, on content moderators, the out-sourced workers who clean up social media feeds, eliminating porn, torture, and other images of political violence. This process he argues is similar to how we read Lacan, who did, after all, refer to the publication of his own books as *poubellication*, punning on the French for garbage, *poubelle.*

Part II of this book admits that our own lack, our lack of knowledge, will always play a role: can we ever know enough about climate change, or does that knowledge constitute an excess, piling up like garbage? In Sasha Langford's "The Psychotopology of Climate," even what appear to be scientific representations, the geometric principles of what she calls "Anthropocene data visualizations," misleadingly contribute to how climate change has accelerated since the early nineteenth century, and do so in a way that flattens—or renders into a two-dimensional line—social effects. In topological models developed by the later Lacan (including the Möbius strip, cross-cap, and Klein bottle) and commented on recently by Žižek, the spatial and the temporal are connected via retroactivity in a dynamic and recursive system that, Langford provocatively argues, show the absential nature of climate change even as they also have made space for the racial and other injustices of an ecological crisis that challenges our measuring systems.[15]

Tackling the entangled relations of subjectivity and science, Juan Luis de la Mora argues, we need a robust, and Lacanian, theory of the subject. In de la Mora's "In Defence of the Subject," he argues that we can best understand the climate crisis via a psychoanalytic theory of affects: of anxiety, enjoyment, and desire. These are the drivers behind our laying waste to the environment, behind our resistance to confronting climate change, behind our fantasy of our own role, and behind what can be our efficacious politics. De la Mora's chapter is especially gripping when he describes the four-fold "depredations" of the contemporary moment: environmental, social, labour, and subjective. And so colonial resource despoliation, epistemic extractivism, the precaritisation of employment, and reticular *dispositifs* simultaneously, via a rich reading of philosophers such as Heidegger and Agamben, alongside Indigenous theory and Lacan, bear witness to obscured problems and solutions.

Calum Matheson's chapter returns to the issue of enjoying the environment in the context of Doomsday Prepping. For Matheson, the enjoyment of the climate emergency can be understood primarily in terms of a *fort-da* repetition that coordinates survivalists' fantasies of the end of the world. Specifically, Matheson argues that rather than produce anxieties that might incite attempts to preserve the earth, these apocalyptic fantasies help distantiate survivalists from the urgency of such tasks. Surprisingly, for Matheson, preppers and popular movies such as *The Day After Tomorrow* illustrate how survivalism and technological optimism are not opposed but rather united insofar as they both share a cultural fantasy of control and the desire to rebuild. Yet such a desire, for Matheson, elides the ultimately horrifying "objective groundlessness of human existence itself" and the knowledge about the possible destruction of the human species in terms of an encounter with the Real.

Part III of the book engages what we call the "Lack of Knowledge." It begins with Alessandra Capperdoni's consideration of what animals want as an enduring question in the philosophical works of Aristotle, Kant, Bataille, Levinas, and others. Acknowledging the enduring question of the animal in philosophic discourse, Capperdoni considers how philosophy is always implicated in the discourse of psychoanalysis. Moving from what philosophy can teach Lacanian psychoanalysis about animals, Capperdoni considers what Lacanian thought might teach philosophy about animals and the non-human, in terms of an "analytic intervention." Capperdoni highlights the importance of ethological research, the decentring of being and thought, and the void that constitutes our (non) being in Lacan's corpus. She also highlights how Lacan's *Seminar I* emphasised the importance of the visual apparatus in humans and nonhumans in establishing the register of the imaginary. She concludes that any claims, Lacanian or otherwise, that tether such voids to a uniquely human experience inevitably instantiate an anthropomorphism that positions the animal as the bearer of the fullness of being.

In Hilda Fernandez-Alvarez's chapter on the "Aokigahara Forest," the idealization of nature, of the ecological, reaches its apex but also its nadir: we now have a natural space for suicides. What, Fernandez-Alvarez asks, can lead Japanese youth to flock to Aokigahara—is there something specific in the social conditions, as a Marxist would argue, is the forest a site for a "dark incandescence" *á la* Bataille, or is the forest akin to the Lacanian letter, an inscription that cuts even as it communicates? It is to Fernandez-Alvarez's credit that she refuses to offer a pat answer, even as she explores what she calls a "psych-ecological" relationship between human subjects and nature. Between the geologic waste described by Lacan and the piles of garbage of the Anthropocene, Fernandez-Alvarez may help us to think of forests as another garbage dump, the perfect site for the ultimate act: suicide.

Surely television advertising is the diametric opposite of refuge in the forest: as Miguel Rivera's chapter on jouissance in TV advertisements argues, however, the difference may be one of degree and not of kind. This argument is pursued via the concept of perversion, and, in particular, Stephanie Swales' influential encapsulating of the Lacanian variation—which should be distinguished from the classic Freudian model.

The perverse SUV advert can safely depict a gas-guzzling vehicle traversing an "unspoiled" landscape, knowing that viewers acknowledge/disavow the contradiction. That is, by Rivera's argument, television advertising constructs its audience not as idiots who cannot see the lies before them, nor as cynics who do not care, but as perverts who, in acts of fetishist disavowal, demonstrate a "both/and" logic. As with Fernandez--Alvarez' locating of her analysis in a field of non-Lacanian theory, Rivera too is able to draw on proximate criticisms in media studies, thereby demonstrating the relevance of psychoanalysis in an expanded context.

Nathan Gorelick situates psychoanalysis at the "end of the world," the title of the book's final Part IV. The chapter begins with a critique of Renee Lertzman's *Environmental Melancholia* insofar as it overlooks the role of enjoyment in the contradictions between people's beliefs about environmental destruction and how their attempts to solve such a problem ends up in facilitating it. Focusing on Hollywood's obsession with the apocalypse, Gorelick addresses the task of moving beyond stultifying anxiety to consider the possibility of an alternative future in which the reconstruction of a new world may commence. Gorelick suggests that the end of the world is easily imagined because it is a very desirable commodity. Moreover, the imaginary allure of the end of the world can be understood in terms of what Matheson calls "messianic-redemptive" and the "naturalistic-sentimental." Such formulas are complementary and comprise two halves of modernity, which from a Lacanian perspective, share the same logic of fantasy.

Matthew Flisfeder's chapter on the Sublime (and the Hysterical Sublime) works through our conflicted relationship to climate disaster via the provocation that we desire the Anthropocene qua obstacle, so we do not have to acknowledge that a pristine nature is not possible. Referring to Fredric Jameson's updating of the Kantian/Burkean sublime to the postmodern, or hysterical, subject, Flisfeder brings into dialogue Lacan's discourse of the hysteric, where a new kind of knowledge is produced, through an encounter with the unconscious. This knowledge—of the capitalist origins of the climate crisis (or what Donna Haraway calls the "Capitalocene")—may be embodied in Jameson's argument elsewhere that postmodernism is the site of the triumph of capital over the last sites of its resistance—the environment and the unconscious. Rather than retreating to recycling and other empty gestures, Flisfeder, ventriloquizing Jameson, suggests we terraform the planet, an appropriately science-fiction conclusion or solution.

Tamás Nagypál's chapter asserts that snowpocalypse films stage an overcoming of the masculine logic of exception via a feminine "counter-ecology." Putting into conversation post-apocalyptic films such as *28 Days Later* and *Mad Max: Road Fury* with Marx's notion of primitive accumulation, Nagypál suggests that these cinematic representations can be best understood as exposing the Capitalocentric underpinnings of the so-called Anthropocene. Nagypál interprets the all-encompassing endeavours of the Capitalocene in terms of the logic of masculine sexuation, which is delimited by a phallic exception. Such a logic is exemplified by Robert Altman's *Quintet* with its portrayal of decay and ruination in a post-apocalyptic future ice age. By contrast, for Nagypál, the recent Chinese blockbuster *The Wandering Earth* uses the ice age to stage an escape from the Capitalocene logic by "problematizing private atmospheres of exception." Nagypál suggests that this manoeuvre is a feminine one because it affirms Bracha L. Ettinger's concept of "matrixial borderlinking" that is capable of overcoming an apocalyptic zero point and reimagining the Anthropocene as a site of shared vulnerability and what Peter Sloterdijk calls "co-immunism."

By way of providing a bravura conclusion to this collection, Todd McGowan argues that self-destruction is *more* natural to the subject than such apparently constitutive activities as reproductive sex or healthy eating. This does not bode well for any politics organised around saving the planet, so it is a good thing that McGowan has given us early warning. When Kant is overcome by the appearance of the stars (a conclusion to the Second Critique which also inspired that famous optimist Samuel Beckett), McGowan argues, he mistakenly turns to moral law as a crutch, when really, he should have engaged in fellatio or eaten a steak. The carnal is the point: sexual and culinary travesties like rape and Big Macs are today explained away with the evolutionary boilerplate of contemporary popularisers who are ignorant of the Lacanian dialectics of the subject: such self-*destruction* leads to self-*reproduction*, for there is no satisfaction like what we gain from our perverse violations of the law.

Notes

1. Sigmund Freud, "A Child Is Being Beaten: A Contribution to the Study of the Origin of Sexual Perversions," in *The Standard Edition of the Complete Psychological Works of Sigmund Freud, Volume XVII (1917–1919): An Infantile Neurosis and Other Works*, trans. James Strachey (London: Hogarth Press and the Institute of Psychoanalysis, 1955), 175–204.
2. For example: "Two of the 14 participants were concerned that increased air temperatures would result in rapid evaporation of the water in their pet bowls, leading to their pets dying of thirst if they didn't check that the water bowls were filled" (Mairwen K. Jones et al., "The Impact of Climate Change on Obsessive Compulsive Checking Concerns," *Australian & New Zealand Journal of Psychiatry* 46, no. 3 [2012]: 268, https://doi.org/10.1177/0004867411433951).
3. Alison Rourke, "Greta Thunberg Responds to Asperger's Critics: 'It's a superpower,'" *Guardian*, September 2, 2019, https://www.theguardian.com/environment/2019/sep/02/greta-thunberg-responds-to-aspergers-critics-its-a-superpower (accessed December 17, 2019).
4. Ilan Kapoor, ed., "The Pervert and the Hysteric: The Politics of Tahrir Square," in *Psychoanalysis and the Global* (Lincoln: University of Nebraska Press, 2018), 257–82.
5. Jacques Lacan, *Écrits: The First Complete Edition in English*, trans. Bruce Fink (New York: W. W. Norton, 2006), 75.
6. Lacan, *Écrits*, 76.
7. *Wikipedia*, s.v. "Slavoj Žižek," last modified May 21, 2020, https://en.wikipedia.org/wiki/Slavoj_Žižek (accessed December 17, 2019).
8. T. J. Demos, "The Agency of Fire: Burning Aesthetics," *e-flux* 98 (February 2019), https://www.e-flux.com/journal/98/256882/the-agency-of-fire-burning-aesthetics/ (accessed December 17, 2019).
9. Marc De Kesel, *Eros and Ethics: Reading Jacques Lacan's Seminar VII*, trans. Sigi Jöttkandt (Albany: SUNY Press, 2009), 225.
10. Jacques Lacan, *The Seminar of Jacques Lacan, Book XI: The Four Fundamental Concepts of Psychoanalysis*, ed. Jacques-Alain Miller, trans. Alan Sheridan (New York: W. W. Norton, 1998), 95; emphasis in the original.
11. Taras Grescoe, *Bottomfeeder: How to Eat Ethically in a World of Vanishing Seafood* (New York: Bloomsbury, 2008).

12. Lacan, *Écrits*, 18.
13. Key discussions of Lacan's sardine can(on) include the following: Joan Copjec, *Read My Desire: Lacan Against the Historicists* (Cambridge, MA: MIT Press, 1994), 31; Todd McGowan, *Psychoanalytic Film Theory and The Rules of the Game* (New York: Bloomsbury, 2015), 74; Henry Krips, "The Politics of the Gaze: Foucault, Lacan and ŽiŽek," *Culture Unbound* 2 (2010): 91–102; and Gautam Basu Thakur and Jonathan Michael Dickstein, *Lacan and the Nonhuman* (London: Palgrave Macmillan, 2018), 5.
14. Jacques Lacan, *The Seminar of Jacques Lacan, Book I: Freud's Papers on Technique, 1953–1954*, ed. Jacques-Alain Miller, trans. John Forrester (New York: W. W. Norton, 1991), 74. Recall Freud on Rome, in his "Civilization and Its Discontents," in *The Standard Edition of the Complete Psychological Works of Sigmund Freud, Volume XXI (1927–1931): The Future of an Illusion, Civilization and its Discontents, and Other Works*, trans. James Strachey (London: Hogarth Press and the Institute of Psychoanalysis, 1961), 57–146.
15. Slavoj Žižek, *Sex and the Failed Absolute* (New York: Bloomsbury, 2019).

Bibliography

Copjec, Joan. 1994. *Read My Desire: Lacan Against the Historicists*. Cambridge, MA: MIT Press.

De Kesel, Marc. 2009. *Eros and Ethics: Reading Jacques Lacan's Seminar VII*. Translated by Sigi Jöttkandt. Albany: SUNY Press.

Demos, T.J. 2019. The Agency of Fire: Burning Aesthetics. *e-flux* 98 (February). Accessed December 17, 2019. https://www.e-flux.com/journal/98/256882/the-agency-of-fire-burning-aesthetics/.

Freud, Sigmund. 1955. A Child Is Being Beaten: A Contribution to the Study of the Origin of Sexual Perversions. In *The Standard Edition of the Complete Psychological Works of Sigmund Freud, Volume XVII (1917–1919): An Infantile Neurosis and Other Works, translated by James Strachey*, 175–204. London: Hogarth Press and the Institute of Psychoanalysis.

———. 1961. Civilization and Its Discontents. In *The Standard Edition of the Complete Psychological Works of Sigmund Freud, Volume XXI (1927–1931): The Future of an Illusion, Civilization and Its Discontents, and Other Works*, translated by James Strachey, 57–146. London: Hogarth Press and the Institute of Psychoanalysis.

Grescoe, Taras. 2008. *Bottomfeeder: How to Eat Ethically in a World of Vanishing Seafood.* New York: Bloomsbury.

Jones, Mairwen K., Bethany M. Wootton, Lisa D. Vaccaro, and Ross G. Menzies. 2012. The Impact of Climate Change on Obsessive Compulsive Checking Concerns. *Australian & New Zealand Journal of Psychiatry* 46 (3): 265–270. https://doi.org/10.1177/0004867411433951.

Kapoor, Ilan. 2018. The Pervert and the Hysteric: The Politics of Tahrir Square. In *Psychoanalysis and the Global*, ed. Ilan Kapoor, 257–282. Lincoln: University of Nebraska Press.

Krips, Henry. 2010. The Politics of the Gaze: Foucault, Lacan and ŽiŽek. *Culture Unbound* 2: 91–102.

Lacan, Jacques. 1991. *The Seminar of Jacques Lacan, Book I: Freud's Papers on Technique, 1953–1954.* Edited by Jacques-Alain Miller. Translated by John Forrester. New York: W. W. Norton.

———. 1998. *The Seminar of Jacques Lacan, Book XI: The Four Fundamental Concepts of Psychoanalysis.* Edited by Jacques-Alain Miller. Translated by Alan Sheridan. New York: W. W. Norton.

———. 1999. *The Seminar of Jacques Lacan, Book XX: On Feminine Sexuality: The Limits of Love and Knowledge.* Edited by Jacques-Alain Miller. Translated by Bruce Fink. New York: W. W. Norton.

———. 2006. *Écrits.* Translated by Bruce Fink. New York: W. W. Norton.

McGowan, Todd. 2015. *Psychoanalytic Film Theory and The Rules of the Game.* New York: Bloomsbury.

Rourke, Alison. 2019. Greta Thunberg Responds to Asperger's Critics: 'It's a Superpower.' *Guardian*, September 2, 2019. Accessed December 17, 2019. https://www.theguardian.com/environment/2019/sep/02/greta-thunberg-responds-to-aspergers-critics-its-a-superpower.

Thakur, Gautam Basu, and Jonathan Michael Dickstein, eds. 2018. *Lacan and the Nonhuman.* London: Palgrave Macmillan.

Wikipedia. 2020. s.v. "Slavoj Žižek." Last modified May 21, 2020. Accessed December 17, 2019. https://en.wikipedia.org/wiki/Slavoj_Žižek

Žižek, Slavoj. 2019. *Sex and the Failed Absolute.* New York: Bloomsbury.

Part I

Lacanian Theory

2

Love Thy Enemy: Environment(al) Politics

Cindy Zeiher

In chronicling the necessary interdependence between two environments, nature and culture, Guy Debord's text, *A Sick Planet* has acquired a new relevance. Whereas Immanuel Kant appealed to the wonder and awe of nature, contemporary culture tends to frame us as sick people inhabiting a sick planet. Thus Debord is to led to critique the counterculture movement of the 1960s and 1970s as one which is obsessed with pollution and its effects[1]:

> Pollution is in fashion today, exactly in the same way as revolution: it dominates the whole life of society, and it is represented in illusory form in the spectacle. It is the subject of mind-numbing chatter in a plethora of erroneous and mystifying writing and speech, yet it really does have everyone by the throat. It is on display everywhere as ideology, yet it is continually gaining ground as a material development. Two antagonistic tendencies, progression towards the highest form of commodity production and the

C. Zeiher (✉)
University of Canterbury, Christchurch, New Zealand
e-mail: cindy.zeiher@canterbury.ac.nz

C. Burnham, P. Kingsbury (eds.), *Lacan and the Environment*, The Palgrave Lacan Series, https://doi.org/10.1007/978-3-030-67205-8_2

> project of its total negation, equally rich in contradictions within themselves, grow ever stronger in parallel with one other. Here are the two sides whereby a sole historical moment, long awaited and often described in advance in partial and inadequate terms, is made manifest: the moment when it becomes impossible for capitalism to carry on working.[2]

According to Debord, we are our own worst enemy because we are destroying the natural environment upon which our social relations depend, indeed the damage is already irreparable and everything we do from now on will be at best palliative. Of course, Debord is stating the obvious but this in turn suggests something more—that ongoing environmental destruction caused by humans is the very thing of narcissistic enjoyment. What he also opens up is the invitation to riff on the word *environment*, even to the extent of perhaps asking the decisive question: *what does the environment want?* This is a question which is essentially one of politics, in particular of who is the neighbour and/or enemy of the environment *qua* nature? Both nature and politics being environments of an extimate space allow us (with a bit of help with Robert Pfaller's theory of interpassivity) ways in which we can think of our relation with the environment, whether natural, political or *politically natural* as not, after all, so sovereign.

For Debord, as for followers of Karl Marx and Louis Althusser, the enemy is the force of capitalism and not just the figure of the capitalist. Here *Das Kapital* is clear: it is the social relations producing and cultivating the condition of class structure which need to be overturned. Thus the real enemy includes environmental destruction which under capitalism is then rewarded via accumulation. This dialectic sets up how the natural environment is treated as a political space of contention, urgency, and continual crisis in which the neighbour/enemy binary occupies a category of its own. Arguably from this point we can consider whether the natural environment is a political or politicised category, whether it is more a matter of agency or rather one of structure. Nature considered as a structureless form does something to the subject in an encounter which intersects space, *jouissance*, and knowledge. The subject may choose to be Kantian, looking to the starry heavens with wonder, humility, and reticence, or like Blaise Pascal, feel simply terrified beyond words. Here the environment is figurative; either our neighbour who comforts us or the enemy who is unpredictable. However, the figures of the neighbour and the enemy are also at the kernel of psychic conflict and are therefore cogent political categories. So intense is this conflict that we consciously mediate it via a strange reconciliation in which conflict is deemed resolvable through the plea for a more nuanced mutual understanding of the neighbour. Moreover, for the subject, conflict is not only extimate but also demands a certain response. Whereas the neighbour, without necessarily commanding any recognition is afforded it in abundance, the enemy, by contrast, demands space even when there is none. Consequently, the figure of the neighbour achieves much greater tenacity within the political environment. We can better understand this through conceptualisation of these two figures. Christ's impossible command to "love thy neighbour as thyself," points towards a collective love inclusive of and extending beyond the self, one which is willing and unconfused, whose uncompromising logic precludes interrogation and doubt because the desire to love emanates from the original big Other in an unquestioned primary conviction and motivation. There is no need to justify the command to love one's neighbour because in so doing one is relying on affective separation from another in which the neighbour becomes a sacred figure. Thus the command to "love one's neighbour" speaks for itself. By contrast and in spite of Christ's exhortation to love and forgiveness of everyone, the enemy remains profane, a threat whose destruction we embrace and enjoy. The figure of the enemy epitomises the adversary whose act of exhibiting hate confounds us, not least through the direct link it affords to our *jouissance*.

Both the neighbour and the enemy exemplify what is at stake regarding the difference between gesture (which implicates the invested figure as an abstraction of meaning) and duty (an obligation to the figure regardless of meaning, role, and function) through their respective metamorphoses of *jouissance* into mere pleasure. In his *On the Pleasure Principal in Culture*, Pfaller's theory of Interpassivity claims that such psychic conflict is exacerbated not necessarily by the creation of the false enemy or the desirable neighbour, but rather by those hidden ideological commands which infiltrate both our private and public lives.[3]

Before we can ascertain who is our neighbour and/or enemy, we need first to consider the context in which these figures are placed. The contemporary context is arguably one we can understand as Pfaller's interpassivity, that is, a context intentionally staged for us and in which we are rendered passive, responsive subjects. In this interpassive environment, the figures of the neighbour and enemy emerge as a form of fetish, easily recognisable as a stand-in for the big Other. Here, interpassivity provides a perfect conduit in which to interrogate the political environment because it offers a way into that which we cannot fully know and are yet complicit. The question then arises, what might be the implications when the demarcation between gesture and duty is no longer clear but blurred or even reversed, for example when the neighbour is afforded mere social gesture rather than social duty?

The figures of neighbour and enemy are social categories, both constructed within the Symbolic Order, which lead to uncertainty and anxiety in the political sphere. They emanate from probing historicisation of the political event and present as signifiers which render the subject paralysed by their passage through politics. The fetish of historicisation can manifest in many ways, for example, as suffocating, conservative nostalgia, fused with humanism regarding both the neighbour and its relation to the enemy. Such a political environment is often mediatised as a universal—the enemy is the inhumane dimension of the neighbour yet both coexist in the same political environment. As Slavoj Žižek argues, the neighbour exists in a neighbourhood which necessarily occupies limits both physical and ideological. That is, subjects inhabit spaces as well as utilize the resources afforded by such spaces. There is an assumption that in inhabiting a space, one has an entitlement to do so. However, as we well know, not everyone inhabits spaces equally and it is this inequality which postulates who is a friend or enemy of the environment, as well as of the neighbourhood. In Georg Hegel's hypothetical and pragmatic ethical community:

> it is easy to say what man must do, what are the duties he has to fulfil in order to be virtuous; he has simply to follow the well-known and explicit rules of his own situation.[4]

To establish an individual ethical stance, even in the context of today's saturated rhetoric, is similarly easy. In the face of conflicting ethical drives, such as climate change denial or giddy environmentalism, the challenge is to achieve an ethical stance which remains anxious and all-knowing in the face of the enemy, because even in the presence of overwhelming authority, one is always grappling with (and enjoying) ever-present lack.

Moreover, the figures of neighbour and of enemy can be understood as inhabiting sovereign moments. In deliberating sovereignty, we must take a step back to consider how this concept has been (and arguably continues to be) a libidinal project. Certainly, Marquis de Sade's work speaks directly to this and in Jacques Lacan's *Écrits,* "Kant *avec* Sade," we witness the literal authoring of sovereignty as a perverse structure: namely that in order to write about sovereignty, one needs to write oneself out of the very process of authorship.[5] Because of Sade's obsession with personal renunciation, we are unable to read his later work which he requested that his son destroy. To find out what Sade has to say about sovereignty, we therefore turn to two important sources: his writing on the libertines and Lacan's reading of Sade's practical logic. For Sade, the libertines desired a kind of sovereignty that deliberately allowed full sovereignty, whereas the other, the slave, was very much dependent on the will-to-sovereignty (arguably, this is Hegel's true freedom). This Sadian fantasy, if one can call it such, is really about returning to structural limits. Here we can say the kernel of sovereignty, to which Lacan alerts us, is the problematic disappearance of the subject through foregrounding the context, which is in this case, environmental. Here we arrive at extimacy.

Lacan's concept of extimacy can be understood through comparison with the Möbius strip whose confusing, spatially intertwined image, sometimes occupying the same space at different times, importantly shares the same edges. It is, claims Lacan, such a confusion which is alienating for the subject. For David Pavón-Cuéllar also, extimacy attributes a spatial status to human desire. Extimacy, "does not simply reside in our outside world, but is the navel, the source of the world, as it is for us."[6] Pavón-Cuéllar orients extimacy towards *Das Ding*, the Thing as a temporal signifier, that which eludes us, that for which we are searching, which is never really an object but rather something else standing in for a named object. *Das Ding* is important for the subject because it to some extent provides the contours of what we are trying to locate, our desire. In the 1968–1969 seminar, *d'un Autre à l'autre*, Lacan formulated such extimacy as *objet a*, the object cause of desire. The secret of *objet a* is that it promises to free the subject from its investment in subjectivity through something beyond subjectivisation, namely an object proper. In *Seminar VII*, Lacan highlights the contradiction that because the elevated object (*Das Ding*) is in part reliant on, as well as concealed in the Imaginary, it is thus a mask which is "the problem of our own action, and sometimes even an alibi, of the effort to focus on the problem of our own action."[7] Lacan goes on to anticipate the *objet a* in his discussion of *Das Ding* as the very thing which is elevated but still retains a source of deep dissatisfaction. It is this dissatisfaction with the elevated object which protects us, but as we later find out, at the same time, it is the *objet a* which is veiling our dissatisfaction. Thus, when one takes into account those guiding notions, ideological conditions and ethical frameworks which appear to govern an environment, be it daily life, its politics, culture, religion, and so on, common-sense injunctions such as inside/outside no longer appear so substantial or easily discernible. To afford such injunctions a priori status is to overlook the shared edges of their constituent parts, to overlook for example, how the body and psyche, culture and language, God and subject coexist in contradiction and even as substitutes for one another. Through the presence of inside and outside in the same place we can perhaps better appreciate the ongoing problem for the divided subject, which is alienation. Lacan insists that alienation is not a mishap instructed by extimacy. Rather, it is an intentional and constitutive feature of being a subject wherein the subject is not whole but split from itself and misrecognised only as fragmented and contradictory in nature. Lacan explicitly states in *Seminar III* that "alienation is constitutive of the imaginary order. Alienation is the imaginary as such."[8]

What might be the relation between such psychoanalytic theorisation to politics when on the face of it, psychoanalysis and politics appear incompatible in occupying the same environment? In classical theory, the primacy of the political is considered to be a general will analogous to liberal society. Here the state reveals itself as a protector in crisis or emergency. Carl Schmitt's *Political Theology* critiques this classical theorisation in suggesting that protecting the social from such crisis presupposes an enemy, reliance on the existence of which is always a precarious way of exercising state power.[9] Schmitt also contends that in creating a political collective, entity or presence, it is not the organisation of bodies which is sufficient for governance but rather the intensification of language. The highly contagious words, "friend or enemy of the state" are for Schmitt a defining feature of liberalism but are also problematic in that they speak to liberalism's covert authoritarianism. The binary, friend or enemy, invokes potential violence and warfare which then become constitutive of politics, for as Schmitt points out there must always be the condition of conflict prior to the conflict itself. Schmitt further contends that democracy justifies itself via a morality propagated through entertainment and economics in which the justification of violence lies in a cause worth fighting for. Consequently, economic and territorial conquests (or losses) are pivoted around the primacy of the political wherein everyone is defined in relation to each other as either friend or foe. For Schmitt, a world without such a defining environment is a world without politics, that is, democracy itself is dependent on the uptake of such categories.

To push this along a bit further, Giorgio Agamben's *Homo Sacer* project approaches "friend or enemy of the state" from a somewhat different angle. For Agamben, sovereign life establishes itself through political order in which the exception (the enemy) provides the law which, having first constituted itself as a rule must "maintain itself in relation to the exception."[10] Thus, for Agamben the state is constituted through a rule of law which both captures and is captured by sovereignty through its very exclusion.

When stripped of the legal status of citizenship and thereby ostracised from the political community this non-citizen, the enemy, is reduced to a biological entity shackled with fault. In this situation the potential for violence escalates. Here, as Lacan observes, the law is complicit with the Thing. So also, as Fredric Jameson reminds us, is language which "can indeed appear within a given historical moment (or shape) as a component of that moment or an event within it."[11] Language itself contributes to historicisation through shaping it in declarations and moral judgements, condemnation and rejection. Žižek attests to this in *Against the Double Blackmail* where the neighbour is projected as just as creepy as the enemy.[12] Although we usually know what the enemy wants—this is generally made clear—the same cannot be said about the desire of the neighbour. As Žižek, drawing on Peter Sloterdijk,[13] points out, politeness forbids asking what the neighbour wants—better to stay out of each other's way as an appropriation of this question. Here discerning alienation can lead to getting work done provided we choose our alienation and use it wisely.

With all this in mind, we can now address Pfaller's theory of interpassivity as one which has the potential to directly formulate extimacy within a logic of politics *qua* nature. For Pfaller, interpassivity delegates enjoyment as an ethical command, notwithstanding this command manifests more as an instructive impression with an object. He offers the excellent example of canned television laughter which acts as a cue so that we know when and how to laugh. This command is instituted for us as an interpassive injunction. Although it has some longevity and even substance, this command is itself a condition formulated through lack and division which in turn sustains not only the command but also the *jouissance* resultant from its being sustained. Pfaller undercuts the taken-for-granted, pseudo-ethical exigencies which underpin the debate of agency and sovereignty, by claiming that the sovereign moment presents us with the almost unbearable and irreconcilable dilemma of not always being able to act in accordance with sovereign agency, a dilemma in which the figure of the neighbour proves a handy substitute for our incapacity. In return, we receive benefits from our socially approved status, living at ease albeit with false consciousness (for some, perhaps not). This also resonates with Joan Copjec's claim in *Read My Desire* that although social spaces are constructed by acts and relations which fill it, they are nevertheless also reduced by such acts and relations.[14] On the face of it, the sovereign moment has agency only when accompanied by the perception that the figure in the act of something obscene is acting authentically, that is, in accordance with both the extimate Symbolic and with other very different rules which accompany it. Therefore, sovereignty is a strange concept to invoke when discussing nature, but its insistence upon the Imaginary provides a seductive paradox: it offers an *experience* but arguably, nothing more. Aaron Schuster rightly asserts that such a sovereign experience affords the subject both speculative and material dimensions:

> It effects a kind of phenomenological reduction of power, reducing it to the realm of pure appearances. Absolute dignity refers to an autonomous, free-floating image, an image of power sovereignly detached from the mundane realities of power, the labor of it worldly exercises.[15]

Is this sovereign experience not also at the very heart of current environmental politics, that collectively, a certain sovereignty is achievable through the political staging of our *oneness* with nature? Moreover, is not this precisely a depiction of the contemporary sovereign subject who is resigned to relinquishing a certain power in favour of passive complicity with the dominant command towards biological salvation through saving the planet? In that this provides unlimited *jouissance*, of course environmentalism makes sense as a political project inclusive of ready-made neighbours and enemies. This is both the sovereign experience and the political platform which signal the hysterical questions: *what does the environment want?, what do I need to be for the environment?* These questions are not so stupid as they might appear, so let us take a step back to consider why. For the hysteric, the point of these questions is not only to confront the Master's command but also the unacknowledged, hidden presuppositions which keep the Master in position. Is not this question (of what we need to be for the environment), precisely what is at play today? If only we could be compliant with one command after another: recycling, reducing air travel, riding a bicycle, eating less meat, buying locally, and so on, then we would be in the position of the hysteric regarding the ultimate *objet a*, this thing called the environment. Such passive uptake of certain activities and prohibitions is precisely what constitutes interpassivity in action. We enjoy not only what is commanded but also how such commands are designated to us.

A helpful way to think about interpassivity is in terms of a visual injunction: both enemy and neighbour are bugs chasing each other on the Möbius strip. Both represent pleasure and dignity and yet are never fully distinguishable to the other, in spite of their mutual perception of being on the other side, continually appearing and disappearing as ethical sovereign moments. The fuller picture includes a supplement in the Symbolic: what the Möbius strip represents is the contextual fantasy in which discourse circulates. We can think of this discourse as political and one which provides the *jouissance* of politics.

In this context, in *Étant donné* (*Being Given*), Jean-Luc Marion's conceptualisation of the neighbour as a moral figure, one who mediates guilt by demanding some sort of civil relation with the enemy, is relevant.[16] Freud identified a basic conflict between the demands of civilised morality and sexual desire: if morality takes the upper hand and the drives are too intense or frustrated to be sublimated, sexuality is either repressed in neurosis or expressed in perverse forms. Here, sovereignty of the neighbour is a place for how the subject might deal with the unresolvable tension of acting in accordance with desire whilst retaining a comprehensible, ethical identity.

Further, Agamben claims that the generalised aura of wartime trauma (we can think of environmental degradation here also) is a common experience of modernity, even a contemporary norm of existence whose negation is a legitimate defensive move in distancing the traumatic event and thereby embracing a new humanity. Here we include not only our *good* neighbour but also any neighbour for whom we atone, for we must attribute to all such figures of exigency an ontological function. It might be argued that Eric Santner's conceptualisation of the neighbour as potentially subversive does not necessarily entail the social rupture it might threaten to the wider political environment.[17] Rather, it interpellates the faithful neighbour as authority's eventual scapegoat willing, if necessary, to be sacrificed for the benefit of civil society. Here the neighbour no longer has a completely positive determination, becoming instead a figurehead in the liberal politics of negation wherein it is a subject of inscription, indeed of enjoyment of the sacrifice we ourselves do not want to make. If the neighbour is thus cast as including the figure of the enemy, it becomes the extimate intruder of our *jouissance*. What does this mean? We know for example, that the enemy for the subject *qua* environmentalist is the climate change denier. However, who is the neighbour within the extimate space, especially when the environmentalist can be both a giddy naturalist enjoying the great outdoors, as well as purport to be a part of a larger collective, by simply *not* being a climate change denier. Might it be that the neighbour is simply whoever does not go along with climate change denial? This simple negation is all too often deemed a sufficient condition for participation in environmental politics wherein the neighbour is merely a figure of social gesture, whereas the enemy one of reified duty, it being my duty not to like my enemy because this represents, as Žižek says, the *inhumane* neighbour.[18] Here the neighbour and enemy, although impenetrable to one and other, are not necessarily strangers in that they are both inscriptions of the political commons.

Should therefore our neighbour trouble us more than our enemy? What comes to mind here is the deeply disturbing line in Wilfred Owen's poem, "Strange Meeting," uttered in an encounter between two soldiers, formerly enemies, after their death in battle: "I am the enemy you killed, my friend."[19] As both neighbours (and enemies) Owen's soldiers harbour a kind of faithless agency to the Symbolic which cannot be reckoned with and in this lies our ultimate impunity. In order to identify with *jouissance*, we are positioned to fight a new and sometimes falsely created enemy by accepting the neighbour as a new, different kind of best friend. In this process the neighbour is depersonalised, has no language and yet is emotionally asserted in the name of (humanist) sovereignty. At the same time and in some strange way we reconstruct, from necessity of the enemy, the neighbour we desire. Such a neighbour represents the threat of not-having and is our defence mechanism. This neighbour is also problematic in that although representing establishment of the sacred, it elicits from us no sacrifice. The enemy demands sacrifice (of ideals), even if this entails the surrender of our conscience, notwithstanding that this might emanate from belief in the superiority of our values. This is precisely what constitutes a problematic for contemporary liberalism: that displaying some semblance of empathy (arguably, the contemporary Master's command) to those who are excluded (that is, enemies who possess neighbourly qualities) is enough for us to posture as emancipatory. Merely wishing for things to be different seems to be a sufficient sovereign moment in which the idea itself presents as enlightening. Yet such a psychic environment harbours unsophisticated and primitive perversity: from a personal position of sympathy I give up nothing yet gain a sanctioned political identity whose apparent cogency ultimately dissolves in vapid politeness. This strangely allows us a way to approach—and hopefully not avoid—the more urgent question of *how do I deal with my enemy as an identification of myself when we occupy the same environment?*

In order to respond to this question we need to think of politics as an environment in itself rather than as merely an aspect of the environment as a whole. Seen this way it would include any environment providing a context which is taken up in the Symbolic Order. However, this adoption, although deliberate, can be only partially grasped because *qua* environment is at best merely a concept, not fully knowable and in any case contentious regarding knowledge and its relation to truth. We need the Imaginary to help fill in the gaps of what we simply do not know about any environment including the shifting and precarious political one. However, we need the category of politics to underpin the discursive environment in which we function and speculate.

Whether the environment we speak of is the natural world or one of politics, it seems what we are always really returning to the classical hysterical question, *what does the environment want?* This question reveals nothing because the very asking of it exposes the limitations of language to alone provide an answer. To be clear, this question concerning the environment is a symptomatic one in that it frames an already occurring apocalyptic event as one whose calamitous effect on both the natural world and politics is finally dawning on us. The question itself has very little to do with what the environment wants from us; the environment is indifferent to the subject. We have become potentially exalted subjects close to extinction; our world is ending and politics reduced to one inane and possibly destructive encounter after another. It would be somewhat naïve and certainly problematic to suppose that looming environmental catastrophe in itself implies a question which can be semantically resolved. Perhaps instead the question, *what does the environment want?* should be deferred, not necessarily because of its apparent impossibility, but rather because it signifies a displacement wherein our desire to do something about the environment exceeds meaning, leaving a void all too readily filled by our will to *jouissance* as unalloyed truth rather than as a state of anguish. The politicisation of the natural environment does a wonderful job of hystericizing the subject. From a void of disinvestment and confronted with the natural environment's passive resistance, the subject hysterically articulates an insistence. Yet we cannot simply replace the category of feminine *joussiance* with that of the environment. This is because first, the appeal to knowing one's *jouissance* is essentially a singular question which has a relation to the phallus, so the question of the environment, which is one of knowing, is problematised if seen as one of retaining *jouissance*. Secondly, the question of the environment presupposes that within every environment there is a self-managing collective force driven by collective consciousness, recognised and harnessed for the purpose of reaching consensus about what the environment wants. Both the giddy naturalist and the idealistic politician are all too ready to tell us what the environment wants and further, how we ought to meet this (Imaginary) command. Here the environment is our symptom ready to enjoy, but because all our questions and knowledge about it are no more than *mi-dire*, we absolutely need to include in these questions the problematic figures of the neighbour and the enemy. Although *mi-dire* connects with knowledge, we are ultimately confronted with the fact that knowledge alone cannot save us from environmental disaster.

Paradoxically the hysteric's question about the environment raises a further more coherent one: *how can I test what I know about this thing called the environment?* This question confronts us with humanity in the Anthropocene whose disturbing political detachment from the environment is tantamount to disavowal. In spite of the *jouissance* emanating from political environmentalism, we continue living in a state of environmental agnosticism. Because our knowledge of the complexity of the environment is always inadequate, we find ourselves in an ambivalent relationship both with the environment and our limited knowledge of it; consequently, we are more at ease when not pressured to know about the findings of environmental science, for example. There is a danger that such ambivalence presents as resignation from which *jouissance* is nevertheless pursued through forms of political engagement, such as supporting our preferred environmental causes, using green energy options, voting Green, recycling, and so on. Yet at the same time the inexorable shrinking of our environmental parameters is forcing us into a revised relationship with that part of the environment we call nature, a relationship which aspires to neatly meet politics in constructing the most ethical, knowledgeable and idolised of all historical figures: the political environmentalist. Is not such misalignment between nature and the environmentalist the very essence of environmental politics? That when it comes to the environment, the Real of politics hits a wall and has nothing to say: all it can do is revere and co-opt the particularities of individuals who obsessively champion their chosen cause. What is being fought for is the relation to a different neighbour, one who fights for the world not to end and whose most obvious enemy is whoever is not on board with this common-sense fight for the politico-ecological. This is the new collective consciousness of consensus, which somersaults political debate by allowing a 16-year-old Swedish girl to speak for the rest of us.[20]

Although the relationship between ourselves and the natural environment plays out against the background of the big Other, at the same time we are (both rightly and unwittingly) deferring the question of what the environment wants, not only through our ambivalence and resignation but also because we see nature as *sujet supposé savoir*. That nature does indeed know is what the procedure of signification implies. Nature presents as an antithetical image of both original sovereign good and latter day Frankensteinan monster created by us. Nostalgia for the former, including memories of a delusory past, inform the illusion that humanity and the environment were once in harmony, yet it is this very mystification of nature which enables the staging of puppet political enemies as monsters, for example climate change deniers, those who use excessive amounts of plastic, purchased consumables produced from exploited labour, and so on. However, these same *monsters* (who are, at the end of the day, comprise of us all) are also telling us something about the Anthropocene which is tough to confront. Even those who simply do not care about the environment or cannot *face the facts* are telling us that the environment not only exists but also that it insists: that is, we are constantly reminded of the potential for history to repeat itself in the worst possible way. Thus the environment draws attention to our dysfunctional relationship with it as well as to *l'différence* which the subject cannot fully grasp. In addition, it speaks to a specific subjective desire which has been historically tracked, namely our desire to control nature and defy mortal existence. When we hear predictions about the end of the world and the extinction of humankind if "we don't do something now," the difference between *now* and *then* is that instead of *then* trying to control nature through technology and exploitation of natural resources, we are *now* declaring our solidarity with nature through bargaining in good faith to find ways of minimising our environmental impact and live more sustainably. Notwithstanding our bad conscience, we are keen to be present, even virtuous, in this absence of knowing. In spite of our apparent disengagement with knowledge of the environment, we attempt to solicit knowledge by offering ourselves to the precarious object that is the environment.

Further, the problem is static emptiness within the term environment seen purely as biosphere *qua* process. This kind of emptiness, or *structurelessness*, as Jo Freeman puts it in her seminal text, is a difficulty not only for the politician but also for the giddy naturalist, who is in need of a Master as political trade-off. For environmental politics "trash" is an unstructured signifier: we are trashing the planet while even green politicians are busy trash-talking each other by co-opting valuable and ejecting dishonourable members arbitrarily.[21] This process is one of ideological *l'différance*, between attempted deferral of the inexorable process of evolutionary change and its Other, the human, exterior process emanating from *l'différance*. Here the environment *consists* of an idea of itself which at the same time insists that it is something which cannot be so easily given up. In so far as the environment functions as an insistent force, one might say that the political environment in which the enemy and the neighbour are most fully constituted, itself aspires to the same condition. Here *l'différance* is a question of what one does with one's trash locates one as enemy or neighbour.

Yet at the same time the environment remains impassive in the face of our concern and endeavour. Consequently, we sometimes feel guilt about our responsibility for environmental destruction, perhaps the most obvious Debordian *jouissance*. To this extent the Anthropocene is not only a concept for thinking or an actuality which is lived, but also a judgement attributing capacity and agency to specific acts which are construed as *urgent* in light of the ecological crisis we face. Here the environment is situated in the realms of both the phenomenological and the political, notwithstanding the opacity of its being judge and jury. Thus the environment is a perfect symptom for us to be grappling with because in pointing towards the tragic nature of existence, it threatens from our sense of awe and wonder.

This symptom gives us a place in the world, a chosen cause upon which to hang our hat. But it is also a signature of life, which registers as conflation between speaking body and signifier, through which we hope to garner knowledge about the environment, if not determine what it wants. Either the environment and its catastrophe are something urgent needing attention now, or it is business as usual as we reconcile with our existence as environmental excess. Either way we are being moved by the environment as an object of *jouissance* and its pure difference with us. By recycling, buying ethically and so on, we oscillate around often glamourous cover-ups of our environmental dysfunction in our compulsion towards obsessional acts. Through such actions we instigate symbolic limits to our relationship with the natural world which nevertheless continually remind us that we are the wrong people living in the wrong world.

The environment is in part constituted as fantasy through which it is confirmed that there is no real relationship with it but rather an incongruous investment in it as *objet a*. In this way we are confronted with an obvious conclusion: we want the environment to provide nothing more than a conduit towards our *jouissance*. However, insofar as the environment has no investment in us, it is indifferent to our *jouissance* which seeks to embody the desire *of* an environment which is necessarily absent from the site of enunciation. Because of the insufficiency of language, the environment retains autonomous propriety over our will to enjoyment, especially in the matter of sustaining our anguish regarding our unknowability. In this way the category of the environment provides for our interpassive jouissance.

The question *what does the environment want?* is very much a reflection of contemporary sentiment rather than of sovereignty because there is neither a clear neighbour to embrace nor a clear enemy to exclude. To be clear, everyone occupies both categories. The universal, complete giddy environmentalist exists no more than does the climate change-denying monster, in so far as these archetypes both speak to Imaginary, absolute Otherness and to the pleasure we gain from attempting to retrieve that which we thought lost, only to realise that we never really had it. The response to *what does the environment want?* lies in its jouissance and inherent unsayability. The environment does not want anything, has never purported to, indeed has never even said anything, and to us, it is indifferent. If we reassign our subjective and political resignation to the realm of impossibility—that is, the environment as *objet a* proper—then we might then surrender to the strange, inevitable silence of a cosmos which owes us nothing.

Notes

1. Written by Guy Debord in 1971, *A Sick Planet*, trans. Donald Nicholson-Smith (London: Bloomsbury, 1971; repr. 2008), was intended for publication in *Internationale Situationniste* 13, but never appeared.
2. Debord, *Sick Planet* (2008), 77.
3. Robert Pfaller, *On the Pleasure Principal in Culture: Illusions Without Owners* (London: Verso, 2014).
4. Georg Wilhelm Friedrich Hegel, *Philosophy of Right*, trans. S. W. Dyde. (London: George Bell, 1896; repr. New York: Prometheus Books; Oxford: Clarendon Press, 1952), § 150.
5. Jacques Lacan, *Écrits: The First complete Edition in English*, trans. Bruce Fink (New York: Norton and Norton, 1966; repr. W. W. Norton, 2002).
6. David Pavón-Cuéllar, "Extimacy," in *Encyclopaedia of Critical Psychology*, ed. Thomas Teo (New York: Springer, 2014), 662.
7. Jacques Lacan, *The Seminar of Jacques Lacan, Book VII: The Ethics of Psychoanalysis, 1969–1960*, ed. Jacques-Alain Miller, trans. Dennis Porter (New York: W. W. Norton; London: Routledge, 1992), 22.
8. Jacques Lacan, *The Seminar of Jacques Lacan, Book III: The Psychoses, 1955–1956*, ed. Jacques-Alain Miller, trans. Russell Grigg (London and New York: Routledge, 1993; repr. 1997), 146.
9. See Carl Schmitt, *Der Begriff des Politischen* (Berlin: Duncker and Humboltm 1976); and *Political Theology: Four Chapter on the Concept of Sovereignty* trans. George Schwab (Chicago: University of Chicago Press, 2006).
10. Giorgio Agamben, *Homo Sacer: Sovereign Power and Bare Life*, trans. Daniel Heller-Roazen (Stanford, CA: Stanford University Press, 1998), 18.
11. Fredric Jameson, "Progress Versus Utopia; or, Can We Imagine the Future?" *Science Fiction Studies* 27, no. 9 (1982): n.p., https://www.depauw.edu/sfs/backissues/27/jameson.html (accessed May 23, 2020).

12. Slavoj Žižek, *Against the Double Blackmail: Refugees, Terror and Other Troubles with Neighbours* (London: Penguin Press, 2016).
13. Slavoj Žižek, *On Violence: Six Sideways Reflections* (New York: St. Martins Press, 2008), 19.
14. Joan Copjec, *Read My Desire: Lacan Against the Historicists* (Cambridge, MA: MIT Press, 1994).
15. Aaron Schuster, "Beyond Satire: The Political Comedy of the Present and the Paradoxes of Authority," in *Sovereignty Inc.: Three Inquiries in Politics and Enjoyment*, ed. William Mazzarella, Eric L. Santner, and Aaron Schuster (Chicago: University of Chicago Press, 2019), 168.
16. Jean-Luc Marion, *Étant donné. Essai d'une phénoménologie de la donation* (Paris: Presses Universitaires de France, 1997); and the English translation, *Being Given: Toward a Phenomenology of Givenness*, trans. Jeffrey L. Kosky (Stanford, CA: Stanford University Press, 2002).
17. See Eric Santner's, "Miracles Happen: Benjamin, Rosenzweig, Freud, and the Matter of the Neighbor," in *The Neighbor: Three Inquiries in Political Theology*, ed. Slavoj Žižek, Eric L. Santner, and Kenneth Reinhard (Chicago: University of Chicago Press, 2013), 76–133; *On Creaturely Life: Rilke, Benjamin, Sebald* (Chicago: University of Chicago Press, 2006); and *The Royal Remains: The People's Two Bodies and the Endgames of Sovereignty* (Chicago: University of Chicago Press, 2011).
18. Slavoj Žižek, "Neighbors and Other Monsters: A Plea for Ethical Violence," in *The Neighbor: Three Inquiries in Political Theology*, ed. Slavoj Žižek, Eric L. Santner, and Kenneth Reinhard, 134–90. Chicago: University of Chicago Press, 2013), 134–90; and Žižek, *Against the Double Blackmail.*
19. Wilfred Owen, "Strange Meeting," in *The Poems of Wilfred Owen*, ed. Jon Stallworthy (New York: W. W. Norton, 1986), 9–10.
20. Greta Thunberg, at the 2019 United Nations' Climate Change Action Summit in New York, said: "You have stolen my dreams and my childhood with your empty words." A more critical and sophisticated response is provided by Slavoj Žižek, "Yes, It Is a Climate Crisis. And Your Tiny Human Efforts Have Never Seemed So Meagre", *Independent*, September 4, 2019, https://www.independent.co.uk/voices/amazon-fires-rainforest-capitalism-bolsonaro-climate-crisis-zizek-a9091966.html (accessed May 23, 2020). This is an interesting comparison, as if Greta's words alone would make a difference to the crisis we are facing (notwithstanding that she is commanding us to act rather than speak), whilst Žižek at least

alerts us to how our tiny human efforts are no more than a virtue-signalling acting out.

21. Jo Freeman, "The Tyranny of Structurelessness," *Berkeley Journal of Sociology* 17 (1972–1973): 151–65.

Bibliography

Agamben, Giorgio. 1998. *Homo Sacer: Sovereign Power and Bare Life*. Translated by Daniel Heller-Roazen. Stanford, CA: Stanford University Press.

Copjec, Joan. 1994. *Read My Desire: Lacan Against the Historicists*. Cambridge, MA: MIT Press.

Debord, Guy. 1971. *A Sick Planet*. Translated by Donald Nicholson-Smith. London: Bloomsbury. Reprint 2008.

Freeman, Jo. 1972–1973. The Tyranny of Structurelessness. *Berkeley Journal of Sociology* 17: 151–165.

Hegel, Georg Wilhelm Friedrich. 1896. *Philosophy of Right*. Translated by S. W. Dyde. London: George Bell. Reprint. New York: Prometheus Books; Oxford: Clarendon Press, 1952.

Jameson, Fredric. 1982. Progress Versus Utopia; or, Can We Imagine the Future? *Science Fiction Studies* 27 (9). Accessed May 23, 2020. https://www.depauw.edu/sfs/backissues/27/jameson.html.

Lacan, Jacques. 1966. *Écrits: The First Complete Edition in English*. Translated by Bruce Fink. New York: Norton and Norton; Reprint. W. W. Norton, 2002.

———. 1968–1969. *Le séminaire XVI: d'un Autre à l'autre*. Paris: Seuil.

———. 1992. *The Seminar of Jacques Lacan, Book VII: The Ethics of Psychoanalysis, 1969–1960*. Edited by Jacques-Alain Miller. Translated by Dennis Porter. New York: W. W. Norton; London: Routledge.

———. 1993. *The Seminar of Jacques Lacan, Book III: The Psychoses, 1955–1956*. Edited by Jacques-Alain Miller. Translated by Russell Grigg. London and New York: Routledge. Reprint 1997.

Marion, Jean-Luc. 1997. *Étant donné. Essai d'une phénoménologie de la donation*. Paris: Presses Universitaires de France.

———. 2002. *Being Given: Toward a Phenomenology of Givenness*. Translated by Jeffrey L. Kosky. Stanford, CA: Stanford University Press.

Owen, Wilfred. 1986. Strange Meeting. In *The Poems of Wilfred Owen*, ed. Jon Stallworthy. New York: W. W. Norton.

Pavón-Cuéllar, David. 2014. Extimacy. In *Encyclopaedia of Critical Psychology*, ed. Thomas Teo, 661–664. New York: Springer.
Pfaller, Robert. 2014. *On the Pleasure Principal in Culture: Illusions Without Owners*. London: Verso.
Santner, Eric L. 2006. *On Creaturely Life: Rilke, Benjamin, Sebald*. Chicago: University of Chicago Press.
———. 2011. *The Royal Remains: The People's Two Bodies and the Endgames of Sovereignty*. Chicago: University of Chicago Press.
———. 2013. Miracles Happen: Benjamin, Rosenzweig, Freud, and the Matter of the Neighbor. In *The Neighbor: Three Inquiries in Political Theology*, ed. Slavoj Žižek, Eric L. Santner, and Kenneth Reinhard, 76–133. Chicago: University of Chicago Press.
Schmitt, Carl. 1976. *Der Begriff des Politischen*. Berlin: Duncker and Humboltm.
———. 2006. *Political Theology: Four Chapter on the Concept of Sovereignty*. Translated by George Schwab. Chicago: University of Chicago Press.
Schuster, Aaron. 2019. Beyond Satire: The Political Comedy of the Present and the Paradoxes of Authority. In *Sovereignty Inc.: Three Inquiries in Politics and Enjoyment*, ed. William Mazzarella, Eric L. Santner, and Aaron Schuster, 161–250. Chicago: University of Chicago Press.
Žižek, Slavoj. 2013. Neighbors and Other Monsters: A Plea for Ethical Violence. In *The Neighbor: Three Inquiries in Political Theology*, ed. Slavoj Žižek, Eric L. Santner, and Kenneth Reinhard, 134–190. Chicago: University of Chicago Press.
———. 2016. *Against the Double Blackmail: Refugees, Terror and Other Troubles with Neighbours*. London: Penguin Press.
———. 2019. Yes, It Is a Climate Crisis. And Your Tiny Human Efforts Have Never Seemed So Meagre. *Independent*, September 4, 2019. Accessed May 23, 2020. https://www.independent.co.uk/voices/amazon-fires-rainforest-capitalism-bolsonaro-climate-crisis-zizek-a9091966.html.

3

"Staying with" the Anxiety: The Ecological *objet petit a* of Inuit Throat Singing

Alois Sieben

A 2019 Canadian national opinion survey on climate change found that "three-quarters of respondents … were worried, with 25% saying they 'think about climate change often and are getting really anxious about it.'"[1] The phrasing of this survey option is slightly ambiguous: is it climate change that the respondents are anxious about, or the fact that they are thinking about it often (and that thought is not really doing anything)? In Timothy Morton's framing of climate change as a hyperobject—meaning an object that is massively distributed across time and space, thus resisting human cognition—he contends that mere thought about climate change is useless; what is required instead is "an affective experience that would existentially and politically bind [humans] to hyperobjects," the most pertinent one being climate change.[2] Yet, perhaps anxiety is too minor or ugly an affect, in theorist Sianne Ngai's words, too suggestive of "obstructed agency" and "passivity" to help save

A. Sieben (✉)
Simon Fraser University, Vancouver, BC, Canada
e-mail: asieben@sfu.ca

C. Burnham, P. Kingsbury (eds.), *Lacan and the Environment*, The Palgrave Lacan Series, https://doi.org/10.1007/978-3-030-67205-8_3

the world.[3] So says Donna Haraway, describing anxiety as the "ineffective" affect marking "the period between about 2000 and 2050 on earth [that will] be called the Great Dithering."[4] In her approach to climate change, Haraway instead advocates for "staying with the trouble," a project involving "generative joy, terror, and collective thinking"[5]; in other words, a rejection of the minor key of anxiety and a return to major key affects alongside the thoughtfulness of Morton's hyper-nightmares.

To stay with the anxiety for a moment, Colette Soler points to its exceptional status among affects according to Jacques Lacan, acting as a crucial bulwark against the subject's total capture by language: "unlike all other affects, [anxiety] is not displaceable" along the signifying chain, but rather "remains moored to what produces it."[6] Anxiety does not deceive, momentarily rescinding the mediating distance of the signifier, and bringing the subject into affective contact with that which it mediates. In this vein, Soler expands on how anxiety indexes a real as it appears in the complicated form of *objet petit a*, allowing us to "get a handle on reality" beyond the signifier's perpetual displacements, and granting it "an epistemological value."[7] Though Lacan's *objet petit a* is similar to Morton's hyperobject in various ways—both are unconscious, extimate, and uncanny—it is different in being immune to any objective capture whatsoever. It is only momentarily translated into the sensation of anxiety, as expressed in Lacan's central formulation from his tenth seminar: anxiety "*is not without an object*," this enigmatic non-object being *objet petit a*.[8] Of note, Lacan allows the attendants of his tenth seminar, *Anxiety*, to comprehend *objet petit a* by passing around a model of the Möbius strip, stating: "this is the *a* … I give it to you much as one might administer the host."[9]

The epistemological anxiety evoked by the *objet petit a* offers a solution to what Bruno Latour describes as "the total disconnect between the range, nature, and scale of the phenomena and the set of emotions, habits of thoughts, and feelings that would be necessary to handle" the "ecological crisis."[10] Latour's proposed solution involves a turn to environmentalist James Lovelock's concept of Gaia, an agential reframing of the environment that can get "rid of us, 'shivering us' out of existence, so to speak."[11] Lovelock's Gaia reconfigures the divide between humanity and the environment to a relationship of "Gaia-in-us" and "us-in-Gaia," bearing a structure that Latour likens to the topology of the Möbius strip.[12] This reference to the Möbius strip is analogous to Lacan's own use of the figure in *Anxiety* to challenge the *Innenwelt* (inner world) and *Umwelt* (environment) dichotomy. In place of this dichotomy, Lacan establishes the Möbius-like *objet petit a* with its affect of anxiety.

Initially, Latour's and Lacan's Möbius strips appear irreconcilable. The Möbius strip proposed by Latour has a base material of organic interchange—between humans and the environment—whereas Lacan's Möbius strip has a base material of symbolic interchange—between the subject and the discursive Other. Along these lines, while Latour justifiably emphasises the need to preserve the natural environment as that which "encompass[es] us,"[13] for Lacan, the human subject lives within the "habitat of the Word that precedes and envelops" them.[14] Nevertheless, this chapter argues that these two Möbius strips overlap according to the structure of the *objet petit a* outlined by Lacan in *Anxiety*. As will be demonstrated through a theoretical excursus into *Anxiety* in the essay's first section, the Möbius-like relationship between the human and the natural environment acts as a pre-existing shadow from which the Möbius-like relationship between the subject and the discursive Other retroactively emerges. To put it more plainly, Lacan's *objet petit a* emerges from—and thus retroactively signals—the discursive Other's supplanting of organic material—or what Lacan calls the "Other environment"[15]—as the habitat of the speaking being. Anxiety is the fleeting sensation of being momentarily suspended between these two environments, at which point, as Dylan Evans explains, the subject "no longer knows where he is."[16]

This theoretical argument will be fleshed out by turning to the music and writing of Inuk multimedia artist Tanya Tagaq, where the line between the speaking being and the organic environment is continually blurred. Tagaq's music is often punctuated by vocal practices adapted from traditional Inuit throat singing, such as heavy or rapid breathing, yelping, moaning, and guttural grunting. These sounds evoke a range of affects in reviewers, from a "sickening intensity,"[17] to an "uncomfortable"[18] feeling, to one reviewer of a Tagaq performance noting: "her sound isn't for everyone: a friend of mine feels anxious and has to leave partway through."[19] On the one hand, Tagaq's singing voice gives body to the primordial nature of the breath, highlighting the human body's natural interchange with oxygen, which is threatened by rising rates of pollution. But on the other hand—that of the symbolic—rather than Tagaq's singing voice embodying the nonconscious breath, her accentuated breath embodies an unconscious voice, or more specifically the voice of the unconscious, described by Lacan as "*this voice which speaks in me, beyond me*."[20] In me, beyond me, in me, beyond me; the unconscious functions similarly to the breath. As will be discussed in the subsequent section, this similarity is essential to the structure of *objet petit a*, being both an organic and symbolic object. The mixing of the organic and the symbolic generates anxiety in the human subject with an ethical resonance, one accomplished in exemplary fashion by Tagaq's art.

The Ecological Foundation of Lacan's *objet petit a*

In order to elucidate the ecological implications of *objet petit a*, it is useful to first examine Lacan's discussion of the subject between two births (real and symbolic) in his tenth seminar, *Anxiety*. Curiously, Lacan's description of birth in this seminar is more complicated than the one he presents in his subsequent eleventh seminar, *The Four Fundamental Concepts of Psychoanalysis*. In the later seminar, Lacan describes the birth of the subject out of the encounter with language in categorical terms, in which "being born with the signifier, the subject is born divided" becoming both "the subject in the field of the drive and the subject as he appears in the field of the Other."[21] As Lacan elaborates, this division of the subject leaves a curious remainder—*objet petit a*—which binds the newly formed relationship between the subject and the Other. Looping like a thread and needle from one piece of fabric to another, the *objet petit a* leaves only rapidly dissolving traces of its handiwork. Through this motion, the subject's inner world and its environment become woven together like the Möbius strip, with the *Innenwelt* emerging out of the *Umwelt* of the Other, meaning the intersubjective field of language. As the price of admission to subjectivity, Lacan details how this Other colonises parts—most infamously, the phallus—or emissions of the subject's body, rendering them as detachable and socially circulatable: signifiers. This is why we can never totally abandon the Other, since we literally have skin in the game. Seeing the body dispatched into the Other in bits and pieces, the subject clutches to objects that might resist signification, like a smelly security blanket. Ultimately, the subject's horror towards its own body being cut up and socially distributed in the Other leads to a questioning of this Other's terrifying desire: *che vuoi* (what do you want)? This perpetual querying—to which there can be no lasting response—establishes a meaningless spot in the Other, a lack.

Lacan's account of the human's birth in *Anxiety* complicates the immaculacy of the symbolic birth. In the earlier seminar, Lacan articulates how the baby "emerge[s] into this world where he must breathe, first and foremost he is literally choked, suffocated" by the trauma of the "inhalation, into oneself, of a fundamentally Other environment."[22] As a reaction to this Other environment, Lacan zeroes in on the human animal's distinctive letting loose of a cry upon being born as the transmission of an "originative anxiety" that appears "prior to any articulation of the Other's demand," this latter Other being the discursive one.[23] Lacan notes how he has "long situated the function of the cry as a relation," as an articulation of demand, this "being the very heart of this [discursive] Other," but this animal cry is something quite different.[24] The "paradox" of this cry for Lacan is how it immediately vanishes upon its appearance, being swallowed by the discursive Other, becoming lost forever; yet, its substance returns, according to Lacan, as an echo in the subject's dialectic with the Other of discourse, repeating itself in a more relational context.[25] As the cry of the baby transforms to the cry of the subject, the original recipient of the cry—the Other environment—transforms to the intersubjective "form of our neighbour."[26] Lacan asserts that this transformation involves the animal being born again with a new Other, akin to "the strangeness of the leap by which living beings left behind their primitive environment and moved over into the air."[27]

From as early as the essay on the mirror stage, Brian Robertson notes how Lacan showed interest in the "state of relative incompleteness with which the human foetus is cast out into the world," drawing largely from anatomist Louis Bolk.[28] Bolk makes the connection between "the prolonged period of the child's biological helplessness and the origin of the family as a cultural institution,"[29] the foundation upon which most subjects encounter language. Lacan suggests that the foetal form of "he [sic] who is going to be the subject"[30] dooms it to being at the mercy of the waiting Other of language, while its cry contains a momentary knowledge—crucially, in the form of anxiety—of its true helplessness. The anxiety expressed in the cry is that of the animal stuck between the jaws of two Others, the Other environment (the real), and the discursive Other. The baby breathes in the Other environment, and breathes out its cry, a cry which Mladen Dolar contends "is immediately seized" by the discursive Other, and "interpreted, endowed with meaning."[31] As Dolar writes, this cry's reception is split between the two Others, being simultaneously a "*cri pur*, a pure scream" and a "*cri pour*, a scream for someone."[32] All anxiety thereafter signals this once felt frontier between the Other environment and the discursive Other.

Though Lacan formalises these two births, the raw material of anxiety that he is working with is largely extracted from Sigmund Freud's work, particularly his 1926 essay "Inhibitions, Symptoms, and Anxiety." As Soler explains, this essay places "at the beginning of mankind's fate the experience of an unmasterable excitation that overwhelms the subject and generates [anxiety] that [Freud] qualifies as 'real,'" later becoming a signal, as both a "memorial of the first trauma and a warning about an imminent danger."[33] Echoing Freud's formulations, Lacan states that "the real … is what anxiety signals."[34] Yet, Lacan diverges from Freud in contending that this real is not that of perpetual alienation due to the intersubjective field of language, or what Soler calls "the bedrock of castration."[35] Instead, Lacan's real involves *objet petit a*, whose affective encounter bears moral implications, indicated by Lacan with the aphorism, "any morality is to be sought on the side of the real."[36] But where might this *objet petit a* be encountered? In Lacan's theory, the human animal is ultimately displaced out of its primordial soup to being a subject within the Other of language, inhabiting a new environment: the "forest of desires"[37] that this Other's lacking structure cultivates. The previous horror of the Other environment becomes bathed in "an ultra-subjective radiance, the foundation of desire"[38]; yet, the *objet petit a* always "lies behind this desire," beyond the tree line, as the limit point of the unnatural radiance of the subject.[39] The *objet petit a* thus lies at the intersection of the territories of the real and the symbolic.

Further delineating the subject's habitat within the symbolic, Lacan notes that while animals can leave traces and even false traces in establishing their territory, human subjects are distinguishable in their capacity to leave "falsely false traces" in what amounts to an attempt to ensure that the Other is forever deceived—it "*musn't know*"—as to the subject's actual territory.[40] For both humans and animals, consequences arise when "something comes and shakes [the] *Umwelt* to its foundations,"[41] such as natural disasters (which of course are on the rise due to climate change). Referencing an "earthquake" or "meteoric mishap," Lacan notes how animals experience an epistemological anxiety and flee in the face of this "manifestation of a locus of the Other … an Other thing [being] evinced as such."[42] Yet, the human subject, whose Other is instead "housed outside of real space," lacks this spatial knowledge, and Lacan suggests would be best served following the animals.[43] The subject's Other cannot appear by any means other than the *objet petit a*, precisely as enigma. While for animals, anxiety is caused by a natural phenomenon like an eclipse of the sun, for the subject, anxiety is caused by a sun suddenly moving in front of a black hole in the sky, an overbearing real presence where there should only be absence. In the face of this sun, the subject experiences the anxiety of a possible melting of the wings of the signifier or the fading of the winds of desire, then the fall towards the "primitive water"[44] of the Other environment from which it was once displaced.

In outlining the subject's falsely false territory, or "home [as] a point located in the Other that lies beyond the image from which we are fashioned,"[45] Lacan owes a further debt to Freud's essay on the uncanny, or the *unheimlich*. Through a linguistic analysis, Freud points out how the *unheimlich* (the unhomely) is paradoxically contained within the multiple valences of *heimlich* (the homely). Like how Lacan's anxiety is not without an object, Freud's *heimlich* is "not unambiguous" in its meaning, being on the one hand "what is familiar and agreeable, and on the other, what is concealed and kept out of sight."[46] Along these lines, Lacan reworks his theory of the mirror stage from the image that we see of ourselves in the mirror being familiar and congenial, to there being "an opening, a gap, where the specular image shows its limit," constituting the "locus of anxiety."[47] This gap is the *objet petit a*, graspable only through the double removes of anxiety and the uncanny. As an example, Freud describes the personal experience of being aboard a train when his compartment door swings open, revealing an elderly gentleman towards whose appearance Freud felt some repellence. Yet, Freud soon realises with dismay that the gentleman is none other than himself, suddenly reflected in the mirror of the swinging door: not himself, but rather not not himself. By losing grip of the mirror image, the subject, being lost, accidentally passes like a stranger into its "lair," and discovers, wallowing furtively there, what Lacan calls "*the object that I am*."[48] The subject thus experiences the feeling described by Freud of "a regression to a time when the ego had not yet marked itself off sharply from the external world."[49] It momentarily grasps its status as part-object in the real, or Other environment. Though Freud develops his concept of the uncanny largely through personal experiences, both he and Lacan contend that the uncanny's "mechanism" is best assembled within art, where its operation can be observed in "less fleeting" fashion.[50] With that in mind, this chapter now turns to the art of Tanya Tagaq, where mechanisms of the uncanny are everywhere, as well as anxiety.

Between Two Others: Tanya Tagaq's Art

Because of settler-colonialism, throat singing was banned during Tanya Tagaq's upbringing in Cambridge Bay, part of over a century-long ban of the practice by Christian missionaries for sounding demonic. Tagaq only encountered it while living in Halifax in her early twenties when she "received a cassette from her mother of two women throat singing."[51] In the process, settler-colonialism transmuted Inuit throat singing into the form of *objet petit a*, evoked by Lacan in the tenth seminar as "the state of a remainder … on the tape reel … of a voice unfastened from its support."[52] Relevantly, Lacan expands this metaphor by placing it in a settler-colonial context. He notes the anxiety displayed in "cultural spheres that we have no reason to call primitive" at the symbolic appropriation of something intrinsic to the body into something "indefinitely reproducible and capable of being spread all over the place," through the invasive mechanisms of the photograph or the gramophone.[53] In this sense, Tagaq's tape cassette is complicated, both a result of settler-colonial unfastening of throat singing from its relational context (both between bodies, and between the body and the land), as well as its mode of survival.

Tagaq's version of throat singing is distinct from the traditional since it is a solo act (in contrast to the traditional pair of competing performers), and more contemporary in its influences and production. Yet, Robyn Taylor-Neu points out that "despite [Tagaq's] avant-garde styling and indie-pop production," reviewers are keen to inscribe her fundamental difference from mainstream music along with her authentic expression of Indigeneity.[54] In this vein, Taylor-Neu situates Tagaq as "a central figure" in "a popularly lauded Indigenous artistic 'renaissance'" in Canada that "rests in part on the capacity of Indigenous authors, visual artists, and performers to 'play Indian' in ways that are legible under the rubric of liberal multiculturalism."[55] In Taylor-Neu's estimation, these artworks are mobilised by the Canadian media in order to prop up a "liberal multicultural fantasy of reconciliation, which operates through an ideal of difference incorporated."[56] Taylor-Neu distinguishes Tagaq as being a singularly potent example of this movement due to her art being primarily expressed through the voice. This voice's singing, wailing, cooing, babbling, and screaming bears the capacity to produce a strong affective encounter in settler audiences, that can be processed or mobilised in the wrong ways. For instance, Taylor-Neu notes how Tagaq's vocal acrobatics are often fantasised as breaking the very limitations of the signifier according to some reviewers, enabling a "genuine communication" between settlers and Indigenous peoples around a shared sense of common humanity beyond language.[57] Following Dene scholar Glen Coulthard's proposal of Indigenous peoples turning away from the liberal politics of recognition, Tagaq's art can be alternatively read as turning away from a relationship with its settler audience towards a different relationship, with the Other environment or what Coulthard describes as "land-connected practices" involving "nonhuman others."[58]

In order to define this alternate relationship, it is useful to turn to Mladen Dolar's conception of the "voice as one of the paramount 'embodiments'" of *objet petit a*.[59] Taylor-Neu quotes a definition from Dolar on the voice as that which "*does not contribute to making sense*."[60]

Yet, as Dolar makes clear, this is only a provisional definition of the voice inherited from the linguistic theories of Roman Jakobson and Ferdinand de Saussure. Rather than merely existing as a beyond to the signifier, Dolar argues that the voice's "proper locus" is at the "'impossible' juncture" of "the signifier [and] the organic," in that it lies at both their cores and "keeps them together."[61] Accordingly, the voice encapsulates the Möbius-like "topology of *objet petit a*."[62] Dolar further circumscribes this locus of the voice through an analysis of the scream, in particular the already discussed infant's cry upon being born. Every scream after this originary one contains the same split function between the *cri pur* upon entering the Other environment and the *cri pour* directed towards the discursive Other. As Dolar writes, the scream "is carried by an interpretation of the unfathomable other with which it tries to cope."[63] A scream is both a non-signifying interpretation of an other (pure enunciation), and an object that demands to be interpreted (as a signifier). The scream exemplifies how "the signifier in general, as such, is possible only as a non-signifier."[64]

As an example from Tagaq's oeuvre of screams, one of the greatest—half a minute in length—came at the end of her performance at the 2014 Polaris Prize award ceremony, where she won best Canadian album with *Animism*. Regarding this album's title, it should be noted that Freud stated that "everything which now strikes us as 'uncanny' fulfils the condition of touching … residues of animistic mental activity within us and bringing them to expression."[65] Freud further clarifies his concept of animism as the belief that "human beings have souls which can leave their habitation and enter into other beings,"[66] that they can traverse to somewhere unhomely. In a similar vein, *Animism* opens with a cover of the Pixies' classic "Caribou." Amid some of the more melodious singing of Tagaq's repertoire, she begins to scream the last word of the lines:

> I live cement, I hate this street
> give dirt to me, I bite lament
> this human form where I was born
> I now repent.[67]

This screamed repentance does not so much illuminate the passage from the subject alienated by the signifier to an animal past, as much as illuminate the impossibility of doing so. In Lacan's terms, her scream "resonates in a void that is the void of the Other,"[68] filling it with sound, and rattling something therein with the pressure of a pure enunciation. To put it differently, her scream fracks the discursive Other. If *Animism* opens with the subject fracking, it closes on the side of the Other undergoing fracking. The final song "Fracking" offers a screaming ventriloquism of the Earth under stress from the eponymous method of resource extraction. As Taylor-Neu notes, some reviewers of *Animism* believe that Tagaq is directly communicating with the real through this ventriloquism, "becom[ing] a mere conduit … rather than authoring each utterance."[69] Yet, the truth to Tagaq's screams lies rather in the middle of these two options—*objet petit a*—as an excess of the real (or the Other environment) into the signifier, and of the signifier into the real. The strength of her screams comes from their inability to be fully heard, either by the Other environment or the discursive Other. It is a scream and nothing more.

If Tagaq's screams are one way in which the gap between the Other environment and the discursive Other is highlighted, her debut novel *Split Tooth* advances a second iteration of this gap through an attempted sexual relationship. The shift from Tagaq's music to her novel is not as drastic as it appears, since Tagaq released an audiobook for *Split Tooth* that contains throat singing interludes between chapters. These interludes recall Lacan's description of feminine or Other *jouissance* in his twentieth seminar, *Encore*, as "*interdit*" (forbidden, just as throat singing was for its own threatening *jouissance*), where he plays on the equivocation of it being *inter-dit*, "said between the words."[70] There is another feminine *jouissance* represented in the text itself through the protagonist having a sexual encounter in which "the backs of [her] eyeballs begin to melt from agony to ecstasy" while her "clit explodes and [she is] split in two from head to toe."[71] From a psychoanalytic perspective, Dylan Evans makes clear that though the sexual relationship suggests a "signification closest to nature in the human being," it is in fact completely entangled with the signifier and thus "there is no such thing, for the human being, as a natural sexual relationship."[72] In the sex of speaking beings, some *jouissance* is always sacrificed. Yet, does this theory completely apply to *Split Tooth*'s representation of a human sexually engaging the Northern Lights? Approaching the protagonist's naked body lying on the ice, the Lights are called "the Masters of Law and Nature" and fill every orifice of her body in a bliss that amounts to "the end of Life."[73] In an image reminiscent of the Möbius strip, "the light from [her] throat joins the light in [her] womb and begins to make a giant fluid figure eight in [her] Body."[74] Her psychic and bodily topology alter, becoming continuous with the Other environment and its Other *jouissance*. In this scene, *Split Tooth* updates Lacan's reference in his twentieth seminar to Gian Lorenzo Bernini's sculpture of the "Ecstasy of Saint Teresa" as an image of feminine *jouissance*.[75] Descending shards of light figure in both, linking the two images.

After this sexual encounter with the Northern Lights, *Split Tooth*'s protagonist notes the psychic change of "calmness replac[ing] anxiety."[76] As Lacan writes, anxiety emanates from the "gap between desire and jouissance,"[77] or what could be alternately described as the gap between the discursive Other and the Other environment. *Split Tooth*'s protagonist traverses anxiety, and establishes a new sort of non-relationship between the speaking being and its natural environment. For those who approach the environment with only desire, such as through the social forces of settler-colonialism, extractive capitalism, and imperialism, the alienness of this *jouissance*, its distance from desire, can only provoke even greater anxieties. Inuk scholar Jackie Price references these material structures of desire with her observation of multiple countries circling the Arctic like vultures, taking part in the discourse that "the Arctic is opening, the Arctic is available, the Arctic is changing, you better get to the Arctic quick if you want access to the Arctic."[78] The sexual enjoyment present in both *Split Tooth* and Tagaq's singing practice—with its similar "heavy breathing and groaning" which is almost "orgasmic" according to reviewer Drew Nelles[79]—conjures a new and terrifying primal scene for the settler audience. Pertinently, overheard "breathing [is] very much part of the first phenomenology of the traumatic scene" according to Lacan, and thus enigmatic respiration forms part of "the original establishing of the function of anxiety."[80] It is through this anxiety that a *jouissance* toxic to the flows of desire damaging the Earth becomes known.

Conclusion

Lacan's initial image of anxiety in his tenth seminar is that of being confronted by a gigantic praying mantis, while wearing a mask that implicates one as either a male or female mantis. As the former, one gets to engage the mantis sexually but also be devoured for one's efforts; as the latter, one is deprived, but survives. Lacan indicates that this mantis is "radically an Other" in that it is the image of an Other outside the Other of intersubjective discourse.[81] Manifesting a similar encounter, Tagaq's art generates an anxiety that places its receivers at the interface of these Others—*objet petit a*—where we truly realise ourselves as unknown objects: we become known unknowns. This anxiety offers the only epistemology capable of locating the human subject momentarily stuck between the real and symbolic, on solid ground, through what Lacan calls anxiety's "grounded certainty."[82] The speaking being can only respond to the ecological crisis from this position, for as Lacan writes, it is "from anxiety that action borrows its certainty. To act is to snatch from anxiety its certainty."[83] Coming at the problem of climate change from the real of anxiety opposes the dangers of a purely symbolic approach, where as Greta Thunberg communicates, "everybody is talking about climate change, but nobody is doing anything,"[84] or a purely imaginary approach, where a fantasy of speaking beings existing in a natural relationship with the Earth predominates. The real of anxiety is the knowledge of our situation on the planet Earth—that it both is and is not our home, and eventually, we must return to it in an uncanny homecoming. Given its worsening conditions, this will likely be a crash landing.

Notes

1. Seth Klein, "When It Comes to Climate Action, the Public Is Ahead of Our Politics: Analysis of National Climate Poll," *Policy Note*, August 12, 2019, http://policynote.ca/climate-poll-2019 (accessed May 24, 2020).
2. Timothy Morton, *Hyperobjects: Philosophy and Ecology after the End of the World* (Minneapolis: University of Minnesota Press, 2013), 184.

3. Sianne Ngai, *Ugly Feelings* (Cambridge, MA: Harvard University Press, 2005), 3.
4. Donna Haraway, *Staying with the Trouble: Making Kin in the Chthulucene* (Durham: Duke UP, 2016), 144.
5. Haraway, *Staying*, 31.
6. Colette Soler, *Lacanian Affects: The Function of Affect in Lacan's Work*, trans. Bruce Fink (London: Routledge, 2016), 19.
7. Soler, *Lacanian Affects*, 20.
8. Jacques Lacan, *The Seminar of Jacques Lacan, Book X: Anxiety, 1962–1963*, ed. Jacques-Alain Miller, trans. A. R. Price (Cambridge: Polity Press, 2014), 89; emphasis in the original.
9. Lacan, *Book X: Anxiety*, 97.
10. Bruno Latour, "Waiting for Gaia. Composing the Common World Through Arts and Politics," lecture, *French Institute*, London, November 2011, 2, http://www.bruno-latour.fr/sites/default/files/124-GAIA-LONDON-SPEAP_0.pdf (accessed May 24, 2020).
11. Latour, "Waiting," 9.
12. Latour, "Waiting," 10.
13. Latour, "Waiting," 9.
14. Soler, *Lacanian Affects*, 54.
15. Lacan, *Book X: Anxiety*, 327.
16. Dylan Evans, *An Introductory Dictionary of Lacanian Psychoanalysis* (New York: Routledge, 1996), 11.
17. Drew Nelles, "Howl: The Rise of Tanya Tagaq," *Walrus*, January 15, 2015, https://thewalrus.ca/howl (accessed May 24, 2020).
18. Julia LeConte, "Tanya Tagaq: Animism," *NOW Toronto*, May 22, 2014, http://nowtoronto.com/music/album-reviews/tanya-tagaq (accessed May 24, 2020).
19. Herb Mathisen, "NOTY Shortlist: The Throat Singer," *Up Here*, December 1, 2015, http://uphere.ca/articles/noty-shortlist-throat-singer (accessed May 24, 2020).
20. Jacques Lacan, *The Seminar of Jacques Lacan, Book II: The Ego in Freud's Theory and in the Technique of Psychoanalysis, 1954–1955*, ed. Jacques-Alain Miller, trans. Sylvana Tomaselli (New York: Cambridge UP, 1988), 171; emphasis in the original.
21. Jacques Lacan, *The Seminar of Jacques Lacan, Book XI: The Four Fundamental Concepts of Psychoanalysis, 1964*, ed. Jacques-Alain Miller, trans. Alan Sheridan (New York: W. W. Norton, 1981), 199.

22. Lacan, *Book X: Anxiety*, 327.
23. Lacan, *Book X: Anxiety*, 326.
24. Lacan, *Book X: Anxiety*, 326.
25. Lacan, *Book X: Anxiety*, 326.
26. Lacan, *Book X: Anxiety*, 326
27. Lacan, *Book X: Anxiety*, 326–27.
28. Brian Robertson, *Lacanian Antiphilosophy and the Problem of Anxiety: An Uncanny Little Object* (London: Palgrave Macmillan, 2015), 173.
29. Robertson, *Lacanian Antiphilosophy*, 174.
30. Lacan, *Book X: Anxiety*, 326
31. Mladen Dolar, *A Voice and Nothing More* (Cambridge, MA: MIT Press, 2006), 27.
32. Dolar, *Voice*, 28.
33. Soler, *Lacanian Affects*, 15.
34. Lacan, *Book X: Anxiety*, 160.
35. Soler, *Lacanian Affects*, 15.
36. Lacan, *Book X: Anxiety*, 148.
37. Jacques Lacan, *The Seminar of Jacques Lacan, Book VII: The Ethics of Psychoanalysis, 1959–1960*, ed. Jacques-Alain Miller, trans. Dennis Porter (New York: W. W. Norton, 1992), 105.
38. Lacan, *Book X: Anxiety*, 219.
39. Lacan, *Book X: Anxiety*, 101.
40. Lacan, *Book X: Anxiety*, 63; emphasis in the original.
41. Lacan, *Book X: Anxiety*, 297.
42. Lacan, *Book X: Anxiety*, 297.
43. Lacan, *Book X: Anxiety*, 297.
44. Lacan, *Book X: Anxiety*, 297.
45. Lacan, *Book X: Anxiety*, 47.
46. Sigmund Freud, *The Standard Edition of the Complete Psychological Works of Sigmund Freud, Volume XVII (1917–1919): An Infantile Neurosis and Other Works*, trans. James Strachey (London: Hogarth Press and the Institute of Psychoanalysis, 1955), 223.
47. Lacan, *Book X: Anxiety*, 108.
48. Lacan, *Book X: Anxiety*, 48; emphasis in the original.
49. Freud, *Works, Volume XVII*, 235.
50. Lacan, *Book X: Anxiety*, 48–49.

51. Laura Stanley, "Tanya Tagaq," *Canadian Encyclopedia*, January 11, 2019, http://thecanadianencyclopedia.ca/en/article/tanya-tagaq (accessed May 24, 2020).
52. Lacan, *Book X: Anxiety*, 274.
53. Lacan, *Book X: Anxiety*, 315.
54. Robyn Taylor-Neu, "'All There Is': The Reconciliatory Poetics of a Singing Voice," *American Anthropologist* 120, no. 1 (March 2018): 113, https://doi.org/10.1111/aman.13003.
55. Taylor-Neu, "All There Is," 113.
56. Taylor-Neu, "All There Is," 114.
57. Taylor-Neu, "All There Is," 118.
58. Glen Coulthard, *Red Skin, White Masks: Rejecting the Colonial Politics of Recognition* (Minneapolis: University of Minnesota Press, 2014), 13.
59. Dolar, *Voice*, 11.
60. Taylor-Neu, "All There Is," 114; emphasis in the original.
61. Dolar, *Voice*, 156–57.
62. Dolar, *Voice*, 73; emphasis in the original.
63. Dolar, *Voice*, 28.
64. Dolar, *Voice*, 29.
65. Freud, *Works, Volume XVII*, 240–41.
66. Sigmund Freud, *Totem and Taboo*, trans. Abraham A. Brill, (New York: Cosimo, 2009), 98–99.
67. Tanya, Tagaq. "Caribou," *Animism* (Toronto: Six Shooter Records, 2014).
68. Lacan, *Book X: Anxiety*, 275.
69. Taylor-Neu, "All There Is," 121.
70. Jacques Lacan, *The Seminar of Jacques Lacan, Book XX: On Feminine Sexuality: The Limits of Love and Knowledge: Encore, 1972–1973*, ed. Jacques-Alain Miller, trans. Bruce Fink, (New York: W.W. Norton, 1999), 119.
71. Tagaq, *Split Tooth*, 113–14.
72. Evans, *Dictionary*, 124.
73. Tagaq, *Split Tooth*, 113–14.
74. Tagaq, *Split Tooth*, 114.
75. Lacan, *Book XX: Encore*, 76.
76. Tagaq, *Split Tooth*, 116.
77. Lacan, *Book X: Anxiety*, 175.

78. Jackie Price, "But You're Inuk, Right?" YouTube video, 1:31:44, lecture, IGOV Indigenous Speakers Series, April 8, 2013, http://youtube.com/watch?v=W36cGxXpjWw (accessed May 24, 2020).
79. Nelles, "Howl."
80. Lacan, *Book X: Anxiety*, 299.
81. Lacan, *Book X: Anxiety*, 325.
82. Lacan, *Book X: Anxiety*, 218.
83. Lacan, Book X: Anxiety, 77.
84. Javier Blas and Laura Millan Lombrana, "Greta Thunberg Tells Davos: You're all talk, no action on climate change," *National Post*, January 21, 2020, http://nationalpost.com/news/world/thunberg-tells-davos-youve-done-nothing-on-climate-change (accessed May 24, 2020).

Bibliography

Blas, Javier, and Laura Millan Lombrana. 2020. Greta Thunberg Tells Davos: You're All Talk, No Action on Climate Change. *National Post*, January 21, 2020. Accessed May 24, 2020. http://nationalpost.com/news/world/thunberg-tells-davos-youve-done-nothing-on-climate-change.

Coulthard, Glen. 2014. *Red Skin, White Masks: Rejecting the Colonial Politics of Recognition*. Minneapolis: University of Minnesota Press.

Dolar, Mladen. 2006. *A Voice and Nothing More*. Cambridge, MA: MIT Press.

Evans, Dylan. 1996. *An Introductory Dictionary of Lacanian Psychoanalysis*. New York: Routledge.

Freud, Sigmund. 1955. *The Standard Edition of the Complete Psychological Works of Sigmund Freud, Volume XVII (1917–1919): An Infantile Neurosis and Other Works*. Translated by James Strachey. London: Hogarth Press and the Institute of Psychoanalysis.

———. 2009. *Totem and Taboo*. Translated by Abraham A. Brill. New York: Cosimo.

Haraway, Donna J. 2016. *Staying with the Trouble: Making Kin in the Chthulucene*. Durham, NC: Duke University Press.

Klein, Seth. 2019. When It Comes to Climate Action, the Public Is Ahead of Our Politics: Analysis of National Climate Poll. *Policy Note*. August 12, 2019. Accessed May 24, 2020. http://policynote.ca/climate-poll-2019.

Lacan, Jacques. 1981. *The Seminar of Jacques Lacan, Book XI: The Four Fundamental Concepts of Psychoanalysis, 1964*. Edited by Jacques-Alain Miller. Translated by Alan Sheridan. New York: W. W. Norton.

———. 1988. *The Seminar of Jacques Lacan, Book II: The Ego in Freud's Theory and in the Technique of Psychoanalysis, 1954–1955*. Edited by Jacques-Alain Miller. Translated by Sylvana Tomaselli. New York: Cambridge University Press.

———. 1992. *The Seminar of Jacques Lacan, Book VII: The Ethics of Psychoanalysis, 1959–1960*. Edited by Jacques-Alain Miller. Translated by Dennis Porter. New York: W. W. Norton.

———. 1999. *The Seminar of Jacques Lacan, Book XX: On Feminine Sexuality, The Limits of Love and Knowledge, Encore, 1972–1973*. Edited by Jacques-Alain Miller. Translated by Bruce Fink. New York: W. W. Norton.

———. 2014. *The Seminar of Jacques Lacan, Book X: Anxiety, 1962–1963*. Edited by Jacques-Alain Miller. Translated by A. R. Price. Cambridge: Polity Press.

Latour, Bruno. 2011. *Waiting for Gaia. Composing the Common World Through Arts and Politics*. Lecture, French Institute, London, November. Accessed May 24, 2020. http://www.bruno-latour.fr/sites/default/files/124-GAIA-LONDON-SPEAP_0.pdf.

LeConte, Julia. 2014. Tanya Tagaq: Animism. *NOW Toronto*. May 22, 2014. Accessed May 24, 2020. http://nowtoronto.com/music/album-reviews/tanya-tagaq.

Mathisen, Herb. 2015. NOTY Shortlist: The Throat Singer. *Up Here*. December 1, 2015. Accessed May 24, 2020. http://uphere.ca/articles/noty-shortlist-throat-singer.

Morton, Timothy. 2013. *Hyperobjects: Philosophy and Ecology after the End of the World*. Minneapolis: University of Minnesota Press.

Nelles, Drew. 2015. Howl: The Rise of Tanya Tagaq. *Walrus*, January 15, 2015. Accessed May 24, 2020. https://thewalrus.ca/howl.

Ngai, Sianne. 2005. *Ugly Feelings*. Cambridge, MA: Harvard University Press.

Price, Jackie. 2013. But You're Inuk, Right? *YouTube Video*. 1:31:44. Lecture, IGOV Indigenous Speakers Series. April 8, 2013. Accessed May 24, 2020. http://youtube.com/watch?v=W36cGxXpjWw.

Robertson, Brian. 2015. *Lacanian Antiphilosophy and the Problem of Anxiety: An Uncanny Little Object*. London: Palgrave Macmillan.

Soler, Colette. 2016. *Lacanian Affects: The Function of Affect in Lacan's Work*. Translated by Bruce Fink. London: Routledge.

Stanley, Laura. 2019. Tanya Tagaq. *Canadian Encyclopedia*. Last Updated January 11, 2019. Accessed May 24, 2020. http://thecanadianencyclopedia.ca/en/article/tanya-tagaq.

Tanya, Tagaq. 2014. *Animism*. Toronto: Six Shooter Records.

———. 2018. *Split Tooth*. New York: Viking.

Taylor-Neu, Robyn. 2018. 'All There Is': The Reconciliatory Poetics of a Singing Voice. *American Anthropologist* 120 (1): 113–125. https://doi.org/10.1111/aman.13003.

4

Confinement and *Jouissance* in Herman Melville's "Bartleby, the Scrivener: A Story of Wall Street"

Alma Krilic

This paper is about Bartleby, the protagonist of Herman Melville's story, and his relationship to his environment. The story is told from the perspective of a lawyer in Manhattan who hires Bartleby as a legal copyist. Soon after he was hired, Bartleby informs his employer that he would "prefer not to" copy or run errands at a moment's notice but curiously remains in the Wall Street office day and night. The lawyer tries everything he could to get Bartleby to work or at least to leave the premises but to no avail.[1]

In this paper, the term *environment* refers to both figurative and literal uses of the word. The former appears in the Lacanian interpretation of Bartleby's famous words "I prefer not to," by Alenka Zupančič, who says that Bartleby's words are void of intent and are not an act of rebellion.[2] Rather, Zupančič uses spatial references to describe the third "space" that is opened up by his words. Zupančič claims that Bartleby's words hover in this third space and blur the distinction between negation and

A. Krilic (✉)
Simon Fraser University, Vancouver, BC, Canada
e-mail: akrilic@sfu.ca

C. Burnham, P. Kingsbury (eds.), *Lacan and the Environment*, The Palgrave Lacan Series, https://doi.org/10.1007/978-3-030-67205-8_4

affirmation. Instead of offering an interpretation of what Bartleby means when he prefers not to work, Zupančič focuses on the form of his words and describes them as an affirmation of negation. Zupančič touches on the topic of the environment in the story but does not specifically address the fact that Bartleby never leaves the office.

Thus, I will continue the discussion by focusing on this element of the story that has been overlooked, which is that Bartleby was confined within the four walls of the Manhattan office throughout the story. I believe that his confinement within the office walls was a symptom of his *jouissance* that contains aggression so strong that the protagonist turns it back upon himself. After all, Bartleby's insistence that he "would prefer not to" is so self-destructive that he starves himself to death.

At first glance, Bartleby's inaction might be interpreted as passivity towards the environment in the broader sense. Bartleby's passivity could be critiqued as an attitude that In *Climate Crisis, Psychoanalysis, and Radical Ethics*, Donna Orange describes as doing nothing or "going about our business as usual:" … "driving to work, flying to conferences, watering our lawns, eating and consuming mindlessly … while human-caused planetary warming threatens to destroy the world."[3] Stated differently, Bartleby's deferral of action can be seen, for instance, as passivity towards the climate crisis as outlined in Orange's book.

Bartleby's preference not to work can also be seen as castration or rendering authority impotent. By preferring not to, Bartleby was not just standing up to the employer's authority and refusing to work. Perhaps Bartleby's "I prefer not to" is a form of passive resistance that renders power and authority impotent as was evident in the boss's futile attempts to get Bartleby to work or to leave the office.

In the following discussion, however, I will neither critique nor praise Bartleby's deferral of action. Rather, I will discuss what lessons can be gleaned by paying attention to his confinement within the Wall Street office. In this instance, I will refer to the environment in a more literal sense. Bartleby's work environment is the material reality of his existence in the lawyer's office, which included not only a partition, chairs and tables, doors and a window, which one would expect in an office environment, but also the presence of other people that would come and go while Bartleby remained in the office. This space also came with implicit expectations that employees are productive and compliant at the office, and leave the office at the end of the workday. I am curious what compelled Bartleby to remain confined in his work environment and transgress the laws and expectations of the Wall Street office, especially since he did not seem to find any purpose or pleasure in being there. Bartleby spent his days and nights at the office, literally never leaving, and in this paper, I claim that Bartleby was neither adapting nor resisting, but rather enjoying his environment.

The Story of Bartleby

The story of Bartleby was written by the American writer Herman Melville in 1853. It first appeared in *Putnam's Magazine* and, in 1856, it was published in Melville's six-story collection titled *The Piazza Tales*. The narrator of the story is a 60-year-old attorney who recently had been promoted. Needing additional help around the law office, he placed an advertisement in the newspaper and soon a "pallidly neat, pitiably respectable, incurably forlorn" figure appeared at the threshold of the office.[4] After a brief conversation and without asking for references, the attorney hires Bartleby on the spot hoping that his sedate character might balance out the fiery tempers of his other employees.

The lawyer assigns Bartleby a desk in the corner of his own office "so as to have this quiet man within easy call, in case any trifling thing was to be done."[5] The arrangement has all the hallmarks of Michel Foucault's description in *Discipline and Punish*, of the disciplinary machinery whose "aim was to establish presences and absences, to know where and how to locate individuals, … to be able at each moment to supervise the conduct of each individual, to assess it, to judge it, to calculate its qualities or merits."[6] By placing the copyist so close to his own desk, the lawyer wanted to ensure that he is the one in control of the office environment and his employee.

The attorney placed Bartleby's desk in the corner next to "a window which originally had afforded a lateral view of certain grimy back-yards and bricks, but which, owing to subsequent erections, commanded at present no view at all, though it gave some light."[7] To complete the arrangement, the lawyer placed between their desks a high green folding screen "which might entirely isolate Bartleby from my sight, though not remove him from my voice. And thus, in a manner, privacy and society were conjoined."[8]

At first, Bartleby "did an extraordinary quantity of writing. As if long famishing for something to copy, he seemed to gorge himself on my documents. There was no pause for digestion. He ran a day and night line, copying by sun-light and by candle-light."[9] From the very beginning, the lawyer observes that Bartleby was always there, writing both by day and night, and there is no mention of where Bartleby lives or what he does when not copying.

But after the first three productive days, Bartleby informs his employer that he would prefer not to check his work for accuracy. I believe that the lawyer's reactions to Bartleby's preference not to work resemble the five stages of mourning developed by the Swiss psychiatrist Elizabeth Kübler-Ross in her 1969 book *On Death and Dying*. Küebler-Ross initially developed what is commonly called the "five stages of grief" as a way to help one deal with one's own imminent death but later expanded it to how people respond to any sort of loss. The five stages consist of denial and isolation, anger, bargaining, depression, and acceptance.[10]

The loss that the lawyer was experiencing was the loss of power over the situation and he thus moves through the stages of grieving his loss of control. After all, he had recently received a promotion and "following his promotion, he had decided to make this person (Bartleby), without objective references, a man of confidence who would owe everything to him. He wants to make him *his* man."[11] And yet, the attorney increasingly felt helpless when confronted with Bartleby's repeated *I prefer not to*.

In my opinion, the first stage of grieving, or denial, is evident when the lawyer hears "I prefer not to" for the first time. The attorney at first denies what he has just heard: "Immediately it occurred to me that my ears had deceived me, or Bartleby had entirely misunderstood my meaning."[12] The lawyer denies the reality of the situation—that he has a scribe who remains in the office constantly writing but preferring not to have his documents checked for accuracy or to run errands at a moment's notice. Disbelief soon turns into anger, and then bargaining that is met with Bartleby's: "At present I would prefer not to be a little reasonable."[13] The lawyer keeps coming up with ways to reach Bartleby and tries everything from reasoning with him to trying to bribe him to leave. The employer's feelings oscillate between empathy and sincere willingness to help the scribe to feelings of resentment and aggravation.

Bartleby never gives an explanation for his odd behaviour or make any demands, making him all the more difficult to negotiate with. The final stage of mourning is when attorney reconciles himself to having around the scribe who soon after declares that he has given up copying altogether. The lawyer tries to run his business as usual and tolerate the scribe who would sit or stand in the corner and stare at the wall. The situation soon causes feelings of resentment and shame in the lawyer who was embarrassed in front of his other clients. In the end, the lawyer accepts the fact that he will never be able to get Bartleby to work or leave the premises and moves his own office, leaving Bartleby behind. Having no alternative, the new occupants of the law office call the police and Bartleby is taken to jail where he dies presumably of hunger—though I believe that it cannot be said with certainty that he starved himself to death. Bartleby could have died, for example, of natural causes or an underlying heart condition but most likely he died because he preferred not to eat.

The Language of Non-Preference

Bartleby's words are not a simple refusal as the brief exchange between the attorney and Bartleby demonstrates. At one point, the attorney asks him: "You *will* not?" to which Bartleby responds: "I prefer not."[14] With the question of "You *will* not?" Zupančič argues that Bartleby's response is not the equivalent to "I don't want to" examine copies, run errands, go home, etc. Rather, it corresponds to "I want to not" examine copies, obey my boss, leave the office, and so on.[15] Similarly, in *The Parallax View*, Slavoj Žižek writes that Bartleby's "I prefer not to" should be taken literally: "It says 'I would prefer not to,' *not* 'I don't prefer (or care) to' … Bartleby does not negate the predicate; rather, he affirms a non-predicate: he does not say that he *doesn't want to do it;* he says that he *prefers (wants) not to do it.*"[16]

Thus, Bartleby's response is not a straightforward refusal. Rather, Bartleby's words are characterised by a lack, and more specifically by a lack of motivation to accomplish anything with "I prefer not to." To illustrate this lack, Zupančič points to the gap between Bartleby's language of non-preference and motivation.[17] The language of preference normally implies emotions, motivation, or engagement. However, there is an absence of any sort of passion in Bartleby as described by the attorney himself: "Had there been the least uneasiness, anger, impatience or impertinence in his manner; in other words, had there been any thing ordinarily human about him, doubtless I should have violently dismissed him from the premises."[18] There seems to be no driving force behind Bartleby's words, no apparent desire to change the existing order of power, and no motivation to outright resist his employer. His words seem devoid of emotion or drive to accomplish anything.

Yet the temptation to look for meaning behind Bartleby's words persists. For example, the attorney in the story falls into this trap when he assumes that Bartleby must know what he means. The attorney asks: "Why do you refuse?",[19] to which Bartleby predictably responds that he would prefer not to respond. Zupančič says that the reader might interpret Bartleby's "I prefer not to" as a metaphor for the lawyer's malaise and boredom with mechanical tasks, and that perhaps Bartleby reflects the lawyer's own desire not to work anymore. Rather than yield to temptation and infuse Bartleby's words with meaning, Zupančič focuses on the form of "I prefer not to" and describes Bartleby's words as an affirmation of negation.

I Prefer Not to as Affirmation of Negation

For Zupančič, Bartleby's formula is not a case of the Freudian *Verneinung*. The function of negation in Freudian psychoanalysis is to bring the content of a repressed thought to the conscious mind of the analysand: "The content of a repressed image or an idea can make its way into consciousness on condition that it is *negated*. Negation is a way of taking cognizance of what is repressed; indeed, it is already a lifting of the repression, though not, of course, an acceptance of what is repressed."[20]

In Freudian psychoanalysis, in order to bring the repressed thoughts to the level of consciousness, the analyst encourages the subject to talk about their dreams and to engage in free associations. The analyst might ask questions, such as "What would you consider the most unlikely imaginable thing in that situation?" or "What do you think was the furthest from your mind at that time?"[21] If the analysand, as Sigmund Freud states "falls into the trap," the answer to such questions is almost always the correct admission. A negation, such as "It's *not* my mother" in my dream,[22] is the ego's recognition of the unconscious, which signals an opening to the unconscious and analysis can begin.

I agree with Zupančič that Bartleby's formula is not a case of negating the existing order of things. The formula does not imply *what* exactly Bartleby is negating, or in other words, the object of his negation is missing. If anything, Bartleby's formula begins with an affirmation *I prefer* and this affirmation has negation itself as its object, *not to*.

Zupančič describes the formula as having a pendulum effect. The sentence swings from affirmation "I would prefer" to the surprising ending in negation "not to," which forces space apart between affirmation and negation. Zupančič refers to this in-between space as a "crack": "or internal interval, that is at work in the relationship between the crucial categorical couples, and that undermines their complementariness and symmetry: inside/outside; pleasure/beyond the pleasure (principle); repression/becoming conscious of the repressed; affective/intellectual; Eros/destructive drive; and so forth."[23]

The formula opens up this third *space* that undermines this complementariness and the symmetry between affirmation and negation. According to Zupančič, Bartleby's words erode the very reality that distinguishes affirmation from negation and the distinction itself is exhausted by this alternative. Thus, Bartleby's formula is not an instance of rebellion and negation of the existing order of things (the boss commands and the clerks work) but rather dwells in this third space, a no-man's land.

Instead of drawing conclusions about Bartleby's words as an affirmation of negation, Zupančič, rather, turns to a discussion about the physical space that Bartleby found himself in.[24] Zupančič conceives of space in Melville's story as a rigid world of work in which space is bound up with certain functions and politics. Everybody knows their place and the tasks that come with it. Once again, this world corresponds best to Foucault's disciplinary machinery where "each individual has his own place; and each place its individual."[25] It is a world of rituals where nothing happens and yet everything changes each time Bartleby utters "I prefer not to." Zupančič says that Bartleby's formula corresponds to another equally famous formula by Mallarme: "Nothing will have taken place but the place itself."[26] For Zupančič, Bartleby is not looking for a new space that will come with new tasks and there is nothing to suggest that Bartleby remained in the office because he needed a roof over his head. Rather, he is looking for a pure place without the symbolic determination, without any politics or functions. If Bartleby wants anything, according to Zupančič, he wants this "Only place itself will have taken place" in Mallarme's formula. He wants place as such, the "placeness" of the place.[27]

Bartleby's Confinement as a Symptom of *Jouissance*

However, Zupančič does not specifically address Bartleby's confinement within this rigid world of work. It may seem that Bartleby ends up in prison at the end of the tale, but it is important to note that he is actually confined throughout the story. His employer notes early on that Bartleby "never went to dinner; indeed he never went anywhere. As yet I had never of my personal knowledge known him to be outside of my office. He was a perpetual sentry in the corner."[28] Only once it seems that Bartleby leaves the office on a Sunday when the employer unexpectedly shows up though it could be that Bartleby was hiding. Bartleby preferred not to leave the office at all, and towards the end he preferred not to eat or even talk to anyone, leading some readers to believe that the protagonist was merely depressed or even anorexic.[29]

I do not think that Bartleby was a depressed anorexic rebel. Like his words, his confinement within the Wall Street office was not an act of defiance. His words seemed void of the intent to rebel against his employer. Similarly, his physical confinement was a gesture of neither adapting nor resisting his environment. Bartleby was not adapting to his work environment by being productive at work and going home. But he did not resist by going on a strike either. Rather, I believe that Bartleby's self-imposed confinement was a symptom of his *jouissance* or a surplus enjoyment generated by his painful existence within the four office walls.

Jouissance or Beyond the Pleasure Principle

Bartleby did not seem to find any pleasure in his confinement. On the contrary, he appeared to be suffering. Bartleby's suffering exemplifies Freud's claim in *Civilization and Its Discontents* that suffering threatens us from mainly three sides: from our own body that is vulnerable to decay, from the external world, and from the social relations within it.[30] It must have been a very painful experience for Bartleby to starve himself to death in isolation. He must have felt lonely, hungry, and humiliated when he was bossed around by the lawyer, ridiculed by visitors, and not to mention being dragged off to jail. Instead of heeding to his boss's pleas to help him, Bartleby does not appear to do anything to avoid this suffering. His persistence shows that perhaps there was a short-circuit in the link between the Freudian pleasure and reality principles that govern "the functioning of our mental apparatus from the start."[31]

The function of the pleasure principle is "to detach from the ego anything that may give rise to such an unpleasurable experience, to expel it and to create an ego that is oriented solely towards pleasure."[32] However, there are limits to this pleasure-oriented ego that needs to learn to distinguish suffering coming from the external or the internal world. For Freud, being able to distinguish between internal suffering and suffering whose origins are in the external world are the ego's first step towards establishing the reality principle which serves a practical purpose to "provide protection against unpleasurable experiences and the threat of them."[33] For example, if I am able to identify the source of suffering as something that is outside of me and in my environment, I might be able to do something to change it and to thereby protect myself from suffering. Perhaps the reality principle was not *kicking in* in Bartleby. He was unable or unwilling to distinguish the source of his suffering as either internal or external, and then try to avoid his suffering.

His persistence points to the possibility that his suffering was generating a surplus that Bartleby began to experience as a pleasurable experience. Jacques Lacan calls this surplus *jouissance* that exhausts Freud's alternative that one feels only pleasure or only pain. Rather, *jouissance* narrows the gap between pleasure and suffering.

In his seminar *The Ethics of Psychoanalysis*, Lacan considers that there is a *jouissance* beyond the pleasure principle that is linked to the drive which compels the subject to transgress the prohibitions imposed on the subject's pursuit of pleasure. In the case of Bartleby, he found pleasure in copying day and night, and never leaving the office.[34] The prohibition on his pursuit of pleasure was embodied in the lawyer's repeated requests to have his work checked for accuracy. Bartleby's "I prefer not to" transgressed his employer's prohibitions and soon contaminated all of his actions, including offerings of food. And starving to death must have been painful. But he persisted. Bartleby seemed to be suffering both spiritually and physically but his persistence points to the possibility that his suffering was tinged with pleasure.

Bartleby's *jouissance,* or this surplus generated by his suffering, has no use value and is not addressed to anyone in particular. Bartleby neither adapts nor resists but rather enjoys his environment by remaining in the office day and night, and engaging in what Melville calls "dead wall reveries," or staring at walls. Such confinement was neither just pleasurable nor only painful. Rather, it was symptom of *jouissance* that is not simply suffering, such as Bartleby's confinement, but is an experience of pain that produces a surplus, or an excess in the form of taking pleasure in one's own pain. And perhaps Bartleby was a masochist.

Jouissance of Transgression

However, this does not account for Bartleby's final act to starve himself to death in confinement. There must have been something in him so aggressive and self-destructive that he felt compelled to commit the final act. This aggressive streak best corresponds to a paradigm of *jouissance* that Lacan calls the "jouissance of transgression."

For Lacan, this paradigm of *jouissance* finds support in transgression of laws that make social existence possible and that govern all forms of social exchange.[35] The Law that governs and makes social life possible is embodied in the commandment "Thou shalt love thy neighbour as thyself," a commandment that Freud himself treated with suspicion. For Freud, the commandment is a principle that is supposed to tame the individual's tendency to aggression: "human beings are not gentle creatures in need of love; on the contrary, they can count a powerful share of aggression among their instinctual endowments."[36] Freud was suspicious of the principle that an individual would love a stranger, their neighbour, the same way they would a family member or a friend. The hesitation to love a stranger is the individual's recognition that the stranger too might harbour the same tendency to aggression and will be tempted to abuse the individual and: "exploit his labour without recompense, use him sexually without his consent, take possession of his goods, humiliate him and cause him pain."[37] The aggressive streak in the individual needs to be tamed because it threatens to disintegrate social life, thus "civilization has to make every effort to limit man's aggressive drives and hold down their manifestations."[38] To make this possible, the Law of loving one's neighbour becomes the organizing principle that supports social exchange and keeps a community together.

Bartleby might give the impression of a passive harmless scribe when in fact his act of confinement was an act of transgression. Bartleby was transgressing the laws of his work environment by confining himself in the Wall Street office, which proved to be so disruptive to the functioning of his law office that the attorney decided to move out, leaving Bartleby behind. His confinement was transgressive because it was disruptive and destructive to both him and to those around him, and it culminated in his arrest.

His confinement was transgressing the laws of the law office and generating a *jouissance* which, for Lacan, finds its support in such transgressions. This particular paradigm of *jouissance* contains an unconscious aggression which the subject does not have immediate access to and which the subject turns back upon himself: "*jouissance* presents itself as buried at the center of a field and has the characteristics of inaccessibility, obscurity and opacity; ... *jouissance* appears not purely and simply as the satisfaction of a need but as the satisfaction of a drive."[39] For Lacan, the *jouissance* of transgression emerges as the satisfaction of not just any drive but the death drive, or aggression.

This aggression might seem to be geared towards the other who poses a threat: "I retreat from loving my neighbour as myself because there is something on the horizon there that is engaged in some form of intolerable cruelty."[40] For Lacan, there is an aggression that persists between the subject and his neighbour and it causes the subject to retreat rather than love the other. This retreat is perhaps best evident in Bartleby's suspicion about the attorney's attempts to love him as his neighbour. Bartleby, rather, retreats not only in his words "I prefer not to" but he also literally retreats into the corner of the law office, becoming immobile. It is in and through this transgression that the death drive is satisfied and Bartleby begins to enjoy his environment. Once he hands himself over to his *jouissance* that contains unconscious aggression and once the process has begun "then there is no longer any limit; it generates ever more powerful aggression in the self,"[41] and Bartleby starves himself to death.

Like his words, Bartleby's confinement was not an act of refusal but rather an act of transgression. Both Bartleby's words "I prefer not to" and his confinement are characterised by a lack that Gilles Deleuze and Jacques Rancière fail to acknowledge when they interpret Bartleby's words as a call for equality.[42] Rather, his words and tranquillity are not a refusal because they lack the intent to resist, to make demands for equality, or to negate the existing order of power in the office. His passivity and his words were acts of transgression because they were not directed at the attorney or at authority in general, lacked use value, and were not addressing anyone in particular. And if his transgressions *were* addressing anything, it would perhaps be the environment that he found himself in. As Zupančič claims, Bartleby wanted the "placeness" of the place,[43] or a pure place without symbolic determination, politics, or functions. Moreover, it is important to also be attentive to his persistence. His confinement must have been an unpleasurable experience and instead of doing something to alleviate suffering, he persists. His persistence points to the possibility that his suffering was generating a surplus that Lacan calls *jouissance.* Perhaps Bartleby was taking pleasure in his pain, retreating "to the unconscious aggression that *jouissance* contains, to the frightening core of the *destrudo*,"[44] and starving to death.

Notes

1. See Herman Melville, “Bartleby, the Scrivener: A Story of Wall-Street,” in *The Piazza Tales* (New York: Dix & Edwards, 1856), 1–29, http://moglen.law.columbia.edu/LCS/bartleby.pdf (accessed May 25, 2020).
2. Alenka Zupančič, “I would prefer not to. 2011,” YouTube video, 1:13:39, February 6, 2013, https://www.youtube.com/watch?v=DYxVmozmScI (accessed May 25, 2020).
3. Donna M. Orange, *Climate Crisis, Psychoanalysis, and Radical Ethics* (London: Routledge, 2017), 11.
4. Mclvillc, “Bartlcby, thc Scrivcncr,” 6.
5. Melville, “Bartleby, the Scrivener,” 6.
6. Michel Foucault, *Discipline and Punish: The Birth of the Prison*, trans. Alan Sheridan (New York: Random House, 1995), 143
7. Melville, “Bartleby, the Scrivener,” 6.
8. Melville, “Bartleby, the Scrivener,” 6.
9. Melville, “Bartleby, the Scrivener,” 6.
10. Elizabeth Küebler-Ross, *On Death and Dying* (New York: Macmillan, 1969).
11. Gilles Deleuze, “Bartleby; or the Formula,” in *Essays Critical and Clinical*, trans. Daniel W. Smith and Michael A. Greco (Minneapolis: University of Minnesota Press, 1997), 75; emphasis in the original.
12. Melville, “Bartleby, the Scrivener,” 7.
13. Melville, “Bartleby, the Scrivener,” 16.
14. Melville, “Bartleby, the Scrivener”; emphasis in the original.
15. Zupančič, “I would prefer not to. 2011.”
16. Slavoj Žižek, *The Parallax View* (Cambridge, MA: MIT Press, 2006), 381; emphasis in the original.
17. Zupančič, “I would prefer not to. 2011.”
18. Melville, “Bartleby, the Scrivener,” 7.
19. Melville, “Bartleby, the Scrivener,” 8.
20. Sigmund Freud, “Negation,” Standard ed., (1925), 235–39, para. 3, emphasis in the original, http://faculty.smu.edu/dfoster/English%203304/Negation.htm (accessed May 25, 2020).

21. Freud, "Negation," para. 2.
22. Freud, "Negation," para. 1; emphasis in the original.
23. Alenka Zupančič, "Not-Mother: On Freud's *Verneinung*," *e-flux Journal* 33 (March 2012): para. 2, https://www.e-flux.com/journal/33/68292/not-mother-on-freud-s-verneinung/ (accessed May 25, 2020).
24. Zupančič, "I would prefer not to. 2011."
25. Foucault, *Discipline and Punish*, 143.
26. Mallarme, quoted by Zupančič, "I would prefer not to. 2011."
27. Mallarme, quoted by Zupančič, "I would prefer not to. 2011."
28. Melville, "Bartleby, the Scrivener," 9.
29. Jane Desmarais, "Preferring not to: The Paradox of Passive Resistance in Herman Melville's 'Bartleby,'" *Journal of the Short Story in English 36* (2001): 25–39, https://journals.openedition.org/jsse/pdf/575 (accessed May 25, 2020).
30. Sigmund Freud, *Civilization and Its Discontents*, trans. David McLintock (London: Penguin Books, 2002), 17.
31. Freud, *Civilization and Its Discontents*, 16.
32. Freud, *Civilization and Its Discontents*, 5.
33. Freud, *Civilization and Its Discontents*, 5.
34. Jacques Lacan, *The Seminar of Jacques Lacan, Book VII: The Ethics of Psychoanalysis, 1969–1960*, ed. Jacques-Alain Miller, trans. Dennis Porter (New York: W. W. Norton; London: Routledge, 1992).
35. Lacan, *Book VII: The Ethics of Psychoanalysis.*
36. Freud, *Civilization and Its Discontents*, 60.
37. Freud, *Civilization and Its Discontents*, 61.
38. Freud, *Civilization and Its Discontents*, 61.
39. Lacan, *Book VII: The Ethics of Psychoanalysis*, 209.
40. Lacan, *Book VII: The Ethics of Psychoanalysis*, 194.
41. Lacan, *Book VII: The Ethics of Psychoanalysis*, 194.
42. See Deleuze, "Bartleby; or the Formula"; and Jacques Rancière, "Deleuze, Bartleby, and the Literary Formula," in *The Flesh of Words: The Politics of Writing*, trans. Charlotte Mandell (Stanford, CA: Stanford University Press, 2004), 146–64.
43. Mallarme, quoted by Zupančič, "I would prefer not to. 2011."
44. Lacan, *Book VII: The Ethics of Psychoanalysis*, 194.

Bibliography

Deleuze, Gilles. 1997. Bartleby; or the Formula. In *Essays Critical and Clinical*, trans. Daniel W. Smith and Michael A. Greco, 68–90. Minneapolis: University of Minnesota Press.

Desmarais, Jane. 2001. Preferring not to: The Paradox of Passive Resistance in Herman Melville's 'Bartleby.' *Journal of the Short Story in English 36*: 25–39. Accessed May 25, 2020. https://journals.openedition.org/jsse/pdf/575.

Foucault, Michel. 1995. *Discipline and Punish: The Birth of the Prison*. Translated by Alan Sheridan. New York: Random House.

Freud, Sigmund. 1925. *Negation*. Standard ed. 235–239. Accessed May 25, 2020. http://faculty.smu.edu/dfoster/English%203304/Negation.htm.

———. 2002. *Civilization and Its Discontents*. Translated by David McLintock. London: Penguin Books.

Kübler-Ross, Elizabeth. 1969. *On Death and Dying*. New York: Macmillan.

Lacan, Jacques. 1992. *The Seminar of Jacques Lacan, Book VII: The Ethics of Psychoanalysis, 1969–1960*. Edited by Jacques-Alain Miller. Translated by Dennis Porter. New York: W. W. Norton; London: Routledge.

Melville, Herman. 1856. Bartleby, the Scrivener: A Story of Wall-Street. In *The Piazza Tales*, 1–29. New York: Dix & Edwards. Accessed May 25, 2020. http://moglen.law.columbia.edu/LCS/bartleby.pdf.

Orange, Donna M. 2017. *Climate Crisis, Psychoanalysis, and Radical Ethics*. London: Routledge.

Rancière, Jacques. 2004. Deleuze, Bartleby, and the Literary Formula. In *The Flesh of Words: The Politics of Writing*, trans. Charlotte Mandell, 146–164. Stanford, CA: Stanford University Press.

Žižek, Slavoj. 2006. *The Parallax View*. Cambridge, MA: MIT Press.

Zupančič, Alenka. 2012. Not-Mother: On Freud's *Verneinung*. *e-flux Journal* 33 (March). Accessed May 25, 2020. https://www.e-flux.com/journal/33/68292/not-mother-on-freud-s-verneinung/.

———. 2013. I Would Prefer Not to 2011. YouTube Video. 1:13:39. February 6, 2013. Accessed May 25, 2020. https://www.youtube.com/watch?v=DYxVmozmScI.

5

Lacan's Trash Talk: Three Objects for the Internet

Clint Burnham

In the 2018 documentary *The Cleaners*, written and directed by Moritz Riesewieck and Hans Block, we meet Filipino "content moderators," the women and men who delete pornographic, violent, and other imagery from social media streams like Facebook, Twitter, and YouTube.[1] In one scene, an employee walks to her home in a slum community beside a garbage dump, saying that her family urged her to get an education so that she would not have to be a garbage picker. Evidently, the film's argument is that content moderation is precisely that sort of work. For, in order to keep people engaged with social media, content moderators pick out the garbage of the internet: dick pics, a painting of Trump with a small penis, beheadings, and other political violence. Content

Thanks especially to Marc Acherman (Langara College), whose discussion of Lacan's three objects, in a PhD qualifying exam almost ten years ago, were very helpful to this chapter. My thanks to Eric Abalajon, who did a sensitivity read of this chapter vis-à-vis its discussion of the Philippines. As always in my writing on Lacan, I have benefited greatly from the collective reading experience of Vancouver's Lacan Salon, since 2007.

C. Burnham (✉)
English Department, Simon Fraser University, Vancouver, BC, Canada
e-mail: clint_burnham@sfu.ca

C. Burnham, P. Kingsbury (eds.), *Lacan and the Environment*, The Palgrave Lacan Series, https://doi.org/10.1007/978-3-030-67205-8_5

moderators work in the "obscene underside" of the internet. However, as the film makes clear, there are different kinds of garbage, different kinds of objects. While videos of bombings in Syria are deleted, those same videos are eagerly sought by human rights groups seeking to document atrocities. Sexual images, like Illma Gore's painting of Trump, which bequeaths upon the President a small penis, may be artistic expression. Political content, with regard to protests in Turkey, may be censored in that country but viewable elsewhere.

In this chapter, I propose to read this digital garbage, these internet objects, in terms of the taxonomy of Jacques Lacan's three objects laid out in the *Encore* seminar. Specifically, I focus on the chapter on "Knowledge and Truth" of *Seminar XX*, where Lacan delineates three objects, the phallic object (Φ, embedded in "scant reality" qua fantasy); S(Ⱥ), or the signifier of the barred Other where truth is never guaranteed (there is no Other of the Other, no metalanguage); and the *objet petit a*, a semblance of being or even its support. Working out the relation of these three objects to digital garbage, that which is deleted by the content moderator (which entails its own taxonomy of negation), this chapter asks: is the internet another example of what Slavoj Žižek calls in his *Enjoy Your Symptom!*, "the inherent imbecility of popular culture," that is, "an *object* in the strict sense of materialized enjoyment—the stain, the uncanny excess that the subjects snatch away from each other"?[2] The garbage of the internet, like the (unconscious) garbage of the social world, is being snatched away by content moderators and other subjects.

To better understand the logic of the three objects, I first look at how Lacan presents the concepts in *Seminar XX*—keeping in mind how much the seminars have their own agenda, which is rarely to clearly elucidate a concept for future university discourses. I then juxtapose a commentary on that *Seminar* by Suzanne Barnard with how Žižek breaks the objects down, first in *Sublime Object of Ideology* and then in *The Plague of Fantasies* (with the added bonus of a footnote in the latter that argues the same "triad of Phi, *a*, and S(Ⱥ), is clearly discernible in the three cinema hits of summer 1996: *Twister, Independence Day, Mission Impossible*)."[3] I then conclude with an algorithmic formula to map out the various subjectivities and objects at work today when we think of garbage, both on and off the internet.

What Is the Garbage in Lacan's Seminars?

It is worthwhile, when discussing Lacan's seminars, to note that the seminars themselves constitute a mass of off-the-cuff comments, looping references to earlier weeks' discussions, bad puns, questions that turn back on the immediately previous statement, which itself may have only been framed in a negative. We find oblique references to literary or theologico-philosophical texts in an intensely metaleptic modality that inevitably, it seems, presents any insight, aphorism, or argument of Lacan's as having already been enunciated. Lacan often and disconcertingly refers to himself in the third person: "Lacan it seems in his first seminar—as it is called—of the year spoke, you'll never guess, about love."[4] An argument we have carefully unpacked may turn out, on a further reading or a different translation or transcription, to be the negation of what one initially read. This could be through a different framing (Lacan often stages, over pages, an argument he is actually discounting), one's own mistake or overly hasty interpretation, or even a realization that this concept is actually at odds with much of what he says elsewhere. Most notoriously, Lacan's programmatic "there is no Other of the Other" turns out to have been pre-emptively contradicted in *Seminar V: Formations of the Unconscious*.[5]

This textual condition may be the result of Lacan's own oral style of presentation, filtered through the varieties of stenographic and tape-recorded transcriptions, the official editing by Jacques-Alain Miller, the bootleg versions and translations.[6] Lacan refers to his *Écrits* as a "poubellication," a pun combining publication and *poubelle*, or garbage can; such a print culture reading further suggests an analogy between that textual condition and the varieties of "data dumps" and actual garbage dumps that are our subject in this chapter.[7] Consider the discussion of "hainamoration" in *Encore*, a Lacanian neologism derived, Fink tells us, from "the noun *haine* ("hate") and the adjective *énamoré* (enamored)."[8] In one version of the seminar, Lacan simply begins "What I will willingly write for you today as '*hainamoration*' is the depth psychoanalysis was able to introduce in order to situate the zone of its experience," in another: "I would like to start from a remark, from a few remarks, the first two make a connection with what today I would be happy to write for you as *hain-amoration*, which must be written h.a.i.n.a.m.o.r.a.t.i.o.n. This as you know, is the relief psychoanalysis was able to introduce in order to situate in it the zone of its experience."[9]

Lacan goes on to criticize the "bastard" term of ambivalence, which no doubt expresses his own ambivalence vis-à-vis Freud, who meant, with the term ambivalence, to suggest a mixture of love and hate which attends any powerful libidinal relation. And if the Miller edits eliminate, or reduce to garbage, the "throat-clearing" with which Lacan introduces the term ("I would like to start from a remark, from a few remarks, the first two make a connection with what today" in Cormac Gallagher's translation above), we have to make two remarks ourselves here. First, note that Miller will nonetheless typically retain such contextualizing or phatic language elsewhere (as in the beginning of the chapter, which includes what is likely a call for questions from the audience: "It would be enough for a hand to go up for me to give that hand the floor, so to speak")[10]; and second, consider whether this editorial scission not only indicates the division of the subject that we can see subventing the structure of garbage but also the logic of which is directly worked out via *hainamoration*. This neologism, *hainamoration*, is a species of *llangue* or lalangue, language permeated by jouissance. *Hainamoration* brings together love and hate more directly than the term ambivalence, and also contains in its very structure a dualism that Lacan is seeking to overturn, in the Seminar as a whole, with his concept of the sexual non-relation.

We will see how Lacan's formula of sexuation relates to his three objects shortly, via Barnard's commentary on *Seminar XX*, but here I want to turn to how Lacan discusses the triangular "graphicisation" ("not to say 'graph,'" he warns us, "because that term has a precise meaning in mathematical logic"!)[11]

This graph (Fig. 5.1) is presented in the session of April 10, 1973, which follows immediately the session in which Lacan presents the formula of sexuation (March 13, 1973). The latter is much more famous and commented upon in the Lacanian corpus, but it is worth noting how, less than a month after devising such a formula, he would continue to offer diagrams to his audience. There is a certain "does not stop being written" (or drawn) quality to Lacan's seminars over the years, with

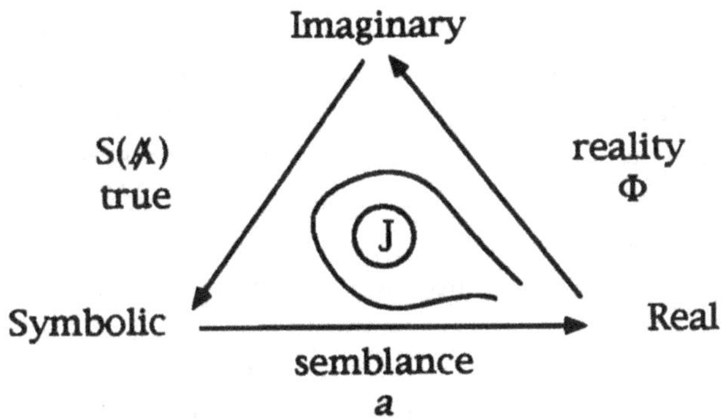

Fig. 5.1 Lacan's jouissance triangle. (Source: https://larvalsubjects.wordpress.com/2008/11/25/sexuation-2-the-logic-of-jouissance/ (accessed May 19, 2020))

diagrams offered on a weekly basis, only a few of which came to have the notoriety of the graphs of desire, the formula of sexuation, or the inverted bouquet. And this question of graphomania will return with the seminar on the *sinthome* a few years later, when Lacan psychoanalyzes Joyce (over-reaching, in my view, and failing to separate sufficiently the author from his text).

But just as Lacan describes in the Fig. 5.1 graph of the three objects, he posits a dialectic of contingency and necessity via the question of "stops not being written" (*cesse de ne pas s'écrire*—the contingent) versus "doesn't stop being written" (*ne cesse pas de s'écrire*—the necessary) and then "doesn't stop not being written" (the impossible).[12] The first formula, "stops not being written," Lacan defines as "the apex of" contingency—that point of analysis, which stops not writing the phallus, or more crucially, produces the S_1, the master signifier. The master signifier now is written, and in its writing, is contingent—it does not matter what letters are used, what phonemes. Garbage is just a word. So is the environment. It stirs up or is a "signifier of the most idiotic jouissance."[13] "Idiotic" also means odd, according to Lacan's Greek etymology; jouissance's idiocy is due to that very contingency. That is, it is stupid to think the word "environment" *means anything*, even while to be an "environmentalist," to belong to the movement called "environmentalism," means believing the word has meaning, is not contingent, is not subjected to the tyranny of the *point de capiton*.

Lacan next engages in a bit of meta-disavowal (it is "from a depreciatory perspective that I contribute the three terms I write as *a*, S(Ⱥ), and Φ"), and then quickly delineates, in reverse order, the meanings of Φ (having to do with a little reality, *peu-de-réalité*) and S(Ⱥ) (asking "what is S(Ⱥ), but the impossibility of telling the whole truth, about which I spoke earlier?") before spending some time on his pet child, the *objet petit a*.[14] Here we find both the *semblant* and the basis or *support* (in the French) of being—which is to say, what we can trace in much of Lacan in terms of the subject as support (or even, to use digital terminology, subject as platform), which derives as much from the Greek *hupokeimenon* and the Latin *subiectum* (via Heidegger) but also and especially as that which *preexists* the subject.[15] That is, our very place as subjects is displaced, we are both the object and the garbage.[16]

As a final remark on Lacan's reading, note that this chapter is named, by Miller, "Knowledge and Truth," which I keep misremembering as "Knowledge and Love." But that is because Lacan will make the very strong argument that the "subject results from the fact that this knowledge must be learned," this knowledge that only exists in the Other, this knowledge that is difficult "less to acquire … than to enjoy," and that the power in knowledge resides in its enjoying.[17] Which brings us again back to our topic, for here we see Lacan acknowledge that "there are things that run and that certainly seem to work like little machines—they are called computers. I am willing to accept the notion that a computer thinks. But that it knows, who would say such a thing? For the foundation of knowledge is that the jouissance of its exercise is the same as that of its acquisition."[18]

What Lacan prompts us then to ask, along with him, is what it means to know. Do we "know" our "knowledge"? Is that what "the jouissance of its [knowledge's?] exercise" means, the enjoyment of knowing? Or does that exercise mean its use, and then how is it different from acquisition? To be sure, learning something is jouissance-filled—it is kind of terrifying, satisfying, and unnerving. When we learn, we often realize how much we still have to learn. We learn what we lack in our knowledge. But why does this complex of jouissance, knowledge, and its exercise/acquisition not apply to computers? And then, what does this question of the computer and jouissance mean in terms of Lacan's three objects? The *true* object, between the Symbolic and the Imaginary, the S(Ⱥ), is the truth of the lack in the Other, the lack of knowledge in the Other—that lack of knowledge in the computer or internet. This is the truth of fake news, or deep fakes, and the like—but also the truth of the lack of knowledge on the part of the content moderators: most notably, in the documentary, when a well-known photograph of US soldiers harassing a prisoner at Abu Ghraib is described as ISIS propaganda! There, content moderators exercise their knowledge, or what they think they know, to make a decision, to delete or block a certain image. This is the meaning, then, of "there is no Other of the Other," no metalanguage: we want our internet to be "clean," but hardly realize that such a standard is neither objective nor realizable, it is rather a matter of our own enjoyment. In like manner, for some subjects, a "clean" environment means cleaning up the Pacific gyre of plastic garbage, while for others it means no garbage in their gated community.[19]

Barnard vs Žižek: *en-corps*

As I noted above, Lacan's diagram of the three objects follows immediately on his formula of sexuation, which has received much more commentary. Barnard's essay, "Tongues of Angels: Feminine Structure and Other Jouissance," first situates Lacan's objects in terms of the sexual non-relation, arguing that while "man is coupled to the Other via object *a*, woman is 'twice' related to the Other—coupled via the phallus and 'tripled' via S(Ⱥ), the signifier of lack in the Other."[20] Barnard adds that jouissance is not so much embodied but a matter of what Lacan calls the "*en-corps*," an enjoying substance that is *beyond* the body, that is a matter of "the death of the body," the "death-in-life attending the 'birth' of the subject," where what is left over from death, "the detritus of meiosis, its 'waste,' returns to haunt the libidinal subject in the form of object *a* ... a 'scrap of the real,' that ex-sists as a residue of a strange form of life."[21] In Barnard's reading of *Encore* (itself owing in some ways to Žižek, whose *Ticklish Subject* and *Metastases of Enjoyment* she cites) *drive and object* "are made of the same stuff," are themselves garbage, the placenta qua maternal Thing.[22] If this formulation means the *objet petit a* is the garbage, the "residual scrap," the foundation or lost object by which man is attracted to the woman, what about the other two objects, Φ and S(A̸)? This is not to argue that Φ and S(A̸) are in any way identical, nor that the trauma of castration results in a break with complementarity, for what is lost is not what is retroactively fantasized as a whole or even as a union of two halves: "what the two sexes lose is not their complementary lost half but an asexual 'sameness,'" and thus the "feminine subject's relation to S(A̸) would produce an 'Other jouissance' related to a different kind of 'knowledge.'"[23] Admittedly, S(A̸), or the signifier of lack in the other, seems like an odd sort of object, and the acknowledgment of the lack in the other seems to trump, in some cases, jouissance as that which we cannot bear.—I use "trump" specifically here because of thinking how that US president, in his blithe lack of interest in the world, surely troubles us more than would malevolent evil. In that way, he constitutes the anal father (as Todd McGowan, drawing on Joan Copjec's comments on Reagan, elaborates), not so much the phallic figure but that of the veiled phallus.[24]

Žižek discusses the phallic object, Phi or Φ, in a few different ways in *Sublime Object* and *Plague of Fantasies*: it is exemplified by the overwhelming birds in Hitchcock, and by the boat at the end of the street in *Marnie*—these material objects that are also maternal objects. They threaten to eat us or swallow us up, to (s)mother us. Such an object, in *The Cleaners*, is the garbage dump in Manila Bay of which we see numerous shots—the actual, existing, material dump next to which we hear that one content moderator say "My mom always told me, that if I don't study well, I'll end up as a scavenger. All they do is pick up garbage. They rely on garbage. … I was afraid of ending up here, picking up garbage. It was one of the reasons that drove me to study well."[25] This garbage dump is also the internet, which our mothers also try to keep us away from, with the difference that the content moderators go through the entire stream of the internet, picking up garbage, rather than where it has been dumped.

And so, then, the garbage dump or the internet, are they the phallic object, Φ? If they are, it is useful to remember, as Žižek points out in *The Plague of Fantasies*, that the three objects themselves serve to *protect* us from jouissance, and that they "specify the three ways in which to 'domesticate' or 'normalize' this horrible Thing … to perceive it in a way which is no longer directly threatening." That is, garbage qua Φ or phallic object protects us from our enjoyment. In *Plague*, Φ is now "the fascinating image which represents the impossible Thing (the *femme fatale* in the *noir* universe, for example)" or the "vortex of the tornado in *Twister*."[26] Φ as overwhelming or fascinating object may be thought of as the ubiquitous images, found online and then deleted by content moderators, of beheadings or suicide bombers, rendered also in Thomas Hirschhorn's video installation "Touching Reality," or in Walid Raad's Atlas Group project *The Thin Neck File*. Matthew Flisfeder has suggested that the cellphone itself is the phallic object *par excellence*,[27] and we can think of the way we hold the phone when we are walking, triumphantly, like an Olympic torch, or old people like myself tilt their heads towards it (to look over their glasses) as if in veneration, or even the way in which we hold the phone for a selfie, our arm hooked as if we are hugging an imaginary friend. If this is so, if the cellphone is Φ, what is it protecting us from?

For Žižek's argument reminds one of nothing so much as Jameson's comments on Walter Benjamin. Jameson is also being psychoanalytic when he argues that Benjamin took Freud's idea that "the function of consciousness is the defense of the organism against shocks from the external environment" and historicized or socialized it, and so "the newspaper acts as a shock absorber," numbing us to novelty but also depriving us of genuine connection between the social world and our private experiences.[28] We can also consider these varieties of the object alongside a similar theory of the object to be found in new materialism, for example Timothy Morton's "hyperobject," which he claims in *Dark Ecology* can be thought of as the Anthropocene itself. Morton flirts with the Anthropocene as unconscious, arguing that the agricultural revolution—he dubs it "agrilogistics"—"had unintentional or unconscious dimensions," and that "ecological awareness is all about having" one's unconscious pointed out.[29] Morton posits, similar to the Lacanian theory of *extimité*, a "refutation of the idea that the unconscious is a region 'below' or 'within' consciousness," and that "we find the unconscious style of a certain mode of human being sprayed all over what lies outside the human, the biosphere. This unconscious," he adds, "is decidedly (*geo*) physical." Morton's tethering of his ideas to a hasty summation of Freud (who he cites here, and

Lacan elsewhere, including the formula for desire, before concluding that, no great surprise, "Lacan was a strongly anthropocentric correlationist") notwithstanding, he does prompt us to wonder if there is an unconscious to the Anthropcene proper.[30] What is disavowed with the concept? Can climate change be the return of the repressed? What is the role of fantasy or enjoyment? Is the Anthropocene the not-all, or is it the primal father? Or, to return to McGowan, the anal father? McGowan does argue that the present day "society of enjoyment" entails a collapsing of the distance from the *objet petit a* (so now, instead of scouring used bookstores for a French copy of *Écrits*, one merely goes onto Amazon—it's just a book you have, instead of a book you desire).[31] In the Anthropocene, nature itself increasingly is that *objet petit a*, the object we cannot attain precisely because we live in the society of enjoyment.

The Internet Is Garbage

By way of summing up or concluding, I would like to offer an algorithm I have been developing lately, that brings together the semiotic rectangle with the formula of sexuation.

In Fig. 5.2, the semiotic rectangle begins with the antinomy of Φ, and the *objet petit a*: two of the three objects, but also opposed because of how they function differently for the feminine and masculine subjects. Then their negations are, respectively, -Φ, and S(Ⱥ), this last entering as the third object of our inquiry. When Φ, and the *objet petit a*, are combined or reconciled, we have the environment, nature, as that impossible phallic place, but also newly our *objet petit a*: what we lack, but somehow imagine as a full presence, the phallic object.[32] Garbage, on the left side, that which combines Φ, and S(Ⱥ), is both that signifier of the lack in the Other and also somehow still the phallic object—in the sense that even, say, when we have all the different "streams" in a bureaucratized recycling station—compost, paper, metal, plastic—we still have "garbage," qua master signifier;[33] but also the way in which the content moderators, removing the "garbage" of obscenity etc. from social media are still working in that "garbage" qua the internet. On the right side, the semblance of being or even its support, its platform or *hypkeimenon* or *subiectum*, denotes both

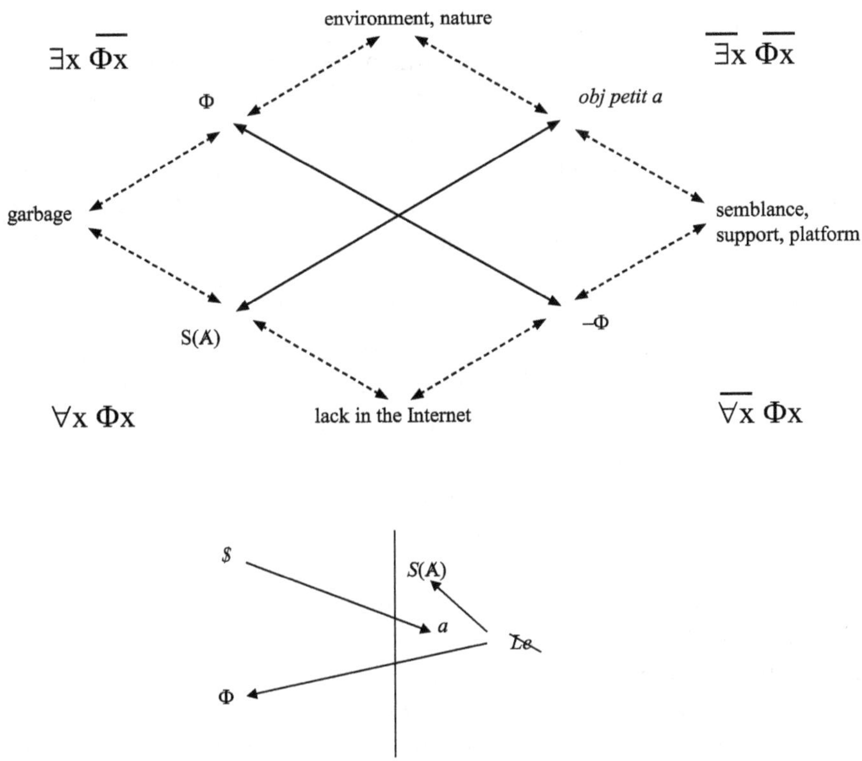

Fig. 5.2 Semiotic rectangle incorporating Lacan's formuala of sexuation. (Source: author)

how Lacan marks the *objet petit a* in his triangular formula but also that which is not phallic—an argument against digital code and platforms as the Real, the actual site for "cleaning up the internet."[34] Which brings us to the bottom of the diagram: the minus-Phi or -Φ, and S(Ⱥ), now signifies the lack in the Internet. The Internet does not know, it is only the *Sujet supposé de Savoir*, only *supposed* to know. Not only because of the fallibility of content moderators (not knowing the difference between ISIS and US torture)—this is perhaps more precisely a matter of *les non-dupes errent*, for really what is the difference, for the human body, for the *homo sacer* of the torture victim?[35]

At the next level, of the four combinations of Lacanian logic, on the lower right and next to the lack in the internet, we have the "not-all," the position of feminine logic, which is a potential lack for the internet, but also suggests how Lacan's texts resist meaning, resist their incorporation as university discourse, to return to this paper's earlier discussion of the status of his seminars.[36] On the upper right, "there does *not* exist anyone not subject to the phallic function": here we have the garbage of the internet qua beheadings, bombings, for there is nothing on the internet that is not subject to imperial cruelty (including the neoliberal outsourcing of its content moderation). On the upper left, the masculine logic of the exception ("there exists one who is *not* subject to castration"), surely describes the recycling subject, who believes he is making a difference. "Every little bit counts!" No: it is precisely every little bit that does *not* count.[37] And on the lower left is "all are subject to the phallic function"—surely now the dick pic (or "vadge" pic) qua uninvited image, but also the cellphone itself, which, even as it assumes near universal global market penetration (!) nonetheless is still castrated, as we saw with the discussion of social media in Myanmar above.

Turning, at last, to the bottom graph of sexuation, $, the masculine subject on the left, here denotes the internet user, who is constantly searching for the *objet petit a* qua Žižek's notion of the MacGuffin, the Real of the symbolic. The barred *Le* replaces the barred *La* in Lacan's formula (the woman who does not exist) with *Le grand Autre*, the big Other that does not exist—the moderator, programmer, or other subject in contradiction to the user. The moderator (the barred *Le*), then, *who instead of searching, picks*, is subject on the one hand to the S(Ⱥ), or "there is no other of the Other," there is no guarantee that an image is or is not pornographic, that something is garbage or not. But the content moderator qua garbage picker is also us, the readers of Lacan (whether in Manila or in Vancouver[38]) encountering Φ, the "throat-clearing" in Lacan, the phatic or metaleptic framing of discourse that obscures whether we are engaging with the Real (a Lacanian soundbite like "a signifier represents a subject to another signifier" or "the unconscious is the discourse of the Other"), and perhaps also denotes the lack in the Anthropocene, or the ways in which that term or concept, along with its proximate terms capitalocene or settlerocene, is always a barred signifier. Combining the *soixante-huit* slogan, "beneath the paving stones, the beach!" with the coder's warning, "garbage in, garbage out," we can conclude that Lacan's lesson is that garbage is what is "in nature more than nature," or "in the Anthropocene more than the Anthropocene" and, indeed, "in Lacan more than Lacan."

Notes

1. Moritz Riesewieck and Hans Block, *The Cleaners* (documentary film) (Berlin: Gebrueder Beetz Filmproduktion et al., 2018). See also "The Trauma Floor: The Secret Lives of Facebook Moderators in America," which provides the following chilling anecdote: "The moderators told me it's a place where the conspiracy videos and memes that they see each day gradually lead them to embrace fringe views. One auditor walks the floor promoting the idea that the Earth is flat. A former employee told me he has begun to question certain aspects of the Holocaust. Another former employee, who told me he has mapped every escape route out of his house and sleeps with a gun at his side, said: 'I no longer believe 9/11 was a terrorist attack.'" My thanks to Sun Ha Hong for this reference. Casey Newton, "The Trauma Floor: The Secret Lives of Facebook Moderators in America," *Verge*, February 25, 2019, https://www.theverge.com/2019/2/25/18229714/cognizant-facebook-content-moderator-interviews-trauma-working-conditions-arizona (accessed July 22, 2019).
2. Slavoj Žižek, *Enjoy Your Symptom! Jacques Lacan in Hollywood and Out* (New York: Routledge, 1992), 22.
3. Žižek, *The Plague of Fantasies* (London: Verso, 1997), 223n.
4. Jacques Lacan, *The Seminar of Jacques Lacan, Book XX: Encore, 1972–1973*, trans. Cormac Gallagher, from unedited French manuscripts (Eastbourne, UK: Antony Rowe, 2010), http://www.lacaninireland.com/web/wp-content/uploads/2010/06/Book-20-Encore.pdf (accessed July 19, 2019).
5. Jacques Lacan, *The Seminar of Jacques Lacan, Book V: Formations of the Unconscious, 1957–1958*, ed. Jacques-Alain Miller, trans. Russell Grigg (Cambridge: Polity Press, 2017), 450–51. See also Lorenzo Chiesa, *Subjectivity and Otherness: A Philosophical Reading of Lacan* (Cambridge: MIT Press, 2007), 104–39.
6. Here, in addition to the Gallagher translation of *Encore* noted above, I also rely on Jacques Lacan, *The Seminar of Jacques Lacan, Book XX: Encore, 1972–1973*, ed. Jacques-Alain Miller, trans. Bruce Fink (New York: Norton, 1998). For a discussion of the editing histories of the Seminars, please see Elisabeth Roudinesco, *Jacques Lacan: A Biography*, trans. Barbara Bray (New York: Columbia University Press, 1999), 529–34.

7. See, for more on *poubellication,* Jean-Claude Milner's intervention in Alain Badiou, *Lacan,* trans. Kenneth Reinhard and Susan Spitzer (New York: Columbia University Press, 2018), 210–211.
8. Lacan, *Encore*, trans. Fink, 90.
9. Lacan, *Encore*, trans. Fink, 90–91; Lacan, *Encore*, trans. Gallagher, 172.
10. Lacan, *Encore*, trans. Fink, 90.
11. Lacan, *Encore*, trans. Fink, 95. Please see also, for a discussion of Lacan's formalization and theories of the state, Anna Kornbluh, "States of Psychoanalysis: Formalization and the Space of the Political." *Theory & Event* 19, no. 3 (2016), https://www.muse.jhu.edu/article/623989 (accessed July 19, 2019); and Suzanne Barnard, "Tongues of Angels: Feminine Structure and Other Jouissance," in *Reading Seminar XX: Lacan's Major Work on Love, Knowledge, and Feminine Sexuality*, ed. Suzanne Barnard and Bruce Fink (Albany: State University of New York, 2002), 171–85.
12. Lacan, *Encore*, trans. Fink, 94.
13. Lacan, *Encore*, trans. Fink, 94.
14. Lacan, *Encore*, trans. Fink, 94–95.
15. In the Greek and Latin, respectively, *hupokeimenon* and *subiectum* constitute the etymologies of the theoretical notion of the subject, both in terms of a basis or platform (*hupokeimenon*) and that which is subjected (*subiectum*). See Barbara Cassin, ed., *Dictionary of Untranslatables: A Philosophical Lexicon.* (Princeton: Princeton University Press, 2014), entry on "subject."
16. Chiesa further breaks down the *objet petit a* into five overlapping functions: as "imaginary representation of lack," as "detachable part-object," as "real lack prior to its imaginarization," as "enigmatic desire of the (m) Other," and as *agalma* (Chiesa, *Subjectivity and Otherness,* 161–63).
17. Lacan, *Encore*, trans. Fink, 96–97. The transferential status of knowledge (akin to Žižek's four "subjects supposed to" know, believe, enjoy, and desire), however, means that the Other merely has to be presumed to know—as Lacan soon says, discussing "Marx and Lenin, Freud and Lacan," that "What is new about their knowledge is that it doesn't presume the Other knows anything about it" (Lacan, *Encore*, trans. Fink, 97). This could be a way of thinking about the S(Ⱥ), or the object as signifier of the lack in the Other.
18. Lacan, *Encore*, trans. Fink, 97.
19. By one account, social media platforms like Facebook are themselves a kind of gated community: This is Virginia Heffernan's argument in

Magic and Loss: The Internet as Art (New York: Simon and Shuster, 2016). See my discussion in *Does the Internet Have an Unconscious? Slavoj Žižek and Digital Culture* (New York: Bloomsbury, 2018), 161, 210. Too, the role of such social media wraps in developing countries (smartphones and watches, tablets, and computers are often sold with Facebook pre-installed), means rapid digital access results in what one human rights group, reporting on Myanmar, called "a crisis of digital literacy: A large population of internet users lacks basic understanding of how to use a browser, how to set up an email address and access an email account, and how to navigate and make judgments on online content" and therefore relies on Facebook as its interface. Please see Business for Social Responsibility (BSR), "Human Rights Impact Assessment: Facebook in Myanmar," October 2018, 12, https://about.fb.com/wp-content/uploads/2018/11/bsr-facebook-myanmar-hria_final.pdf (accessed May 27, 2020).

20. Barnard, "Tongues of Angels," 172.
21. Barnard, "Tongues of Angels," 174. Here Barnard signals a key difference between Lacanian theory and some affect and feminist theory today, which argues for the importance of "embodied" knowledges (see, for example, the "queer phenomenology" of Sara Ahmed).
22. Barnard, "Tongues of Angels," 175.
23. Barnard, "Tongues of Angels," 176.
24. See, for comments on leaders as anal fathers (akin to Žižek's analysis of the "master with the pants down"), Todd McGowan, *The End of Dissatisfaction? Jacques Lacan and the Emerging Society of Enjoyment* (Albany: State University of New York, 2004), 53–54. McGowan references Joan Copjec, who is worth quoting here: "America's solution is, in analytic terms, hysterical: one elects a master who is demonstrably fallible—even, in some cases, incompetent. What may first appear to be a stumbling block turns out on closer inspection to be a solution: Americans love their masters not simply in spite of their frailties but because of them" in *Read My Desire: Lacan Against the Historicists* (Cambridge: MIT Press, 1994), 149. Can this not be said of garbage as well, be it digital (virtual) or actual? Let alone Trump. McGowan elaborates on how we submit not to the phallus but its veil in his recent essay on Lacan's "Signification of the Phallus": "a subject acts in thrall to the phallus not when displaying what appears as clearly phallic behavior—like a muscled Arnold Schwarzenegger evincing indestructibility in

The Terminator ... but instead when a seemingly innocuous figure exercises an authority that doesn't appear as authority" in *Reading Lacan's* Écrits: *From* Signification of the Phallus *to* Metaphor of the Subject, ed. Stijn Vanheule, Derek Hook, and Calum Neill (London: Routledge, 2018), 14. That is, recycling is actually garbage.

25. Moritz Riesewieck and Hans Block, *The Cleaners*. Eric Abalajon notes, regarding this moment in the documentary: "This is not simple upward social mobility aspirations, but one that is built on a colonial past and a neoliberal present. Education system here is still modeled to cater needs of capitalist metropoles. Primarily, you either go out to become a care worker (nurse, caregiver, nanny, cleaning lady, etc) or you stay and work in satellite offices of foreign capital—of which business processing outsourcing (BPOs) are the latest players. Filipino workers are fluent in English, computer-literate, and wages are cheap. You study hard, but in a sense you still end up doing the empire's dirty jobs." Email, 17 July, 2020.
26. Žižek, *Plague of Fantasies*, 175–76.
27. Matthew Flisfeder, personal communication with the author, email, 26 October, 2018. A Simon Fraser University doctoral student, Rawia Inaim, has also made the same remark about the role of the cellphone in a sex scene between Frank Underwood and Zoe Barnes in the first season of *House of Cards*, retroactively heightened by accusations of sexual predation against Kevin Spacey, who plays Underwood.
28. Fredric Jameson, "Walter Benjamin, or Nostalgia," *Salmagundi* 10–11 (1969–1970): 54–55, https://www.jstor.org/stable/40546514.
29. Timothy Morton, *Dark Ecology: For a Logic of Future Coexistence* (New York: Columbia University Press, 2016), 23.
30. Morton, *Dark Ecology*, 122, 127.
31. McGowan, *End of Dissatisfaction?*, 75–80.
32. I devised an earlier form of this algorithm with garbage and the environment/nature as antinomic. But that does not work: we do not actually have access to those forms; they are only what we get when we *combine* Φ, and the *objet petit a* (nature) or when we *combine* Φ, and S($\cancel{A}$) (garbage)—that is, nature or garbage are not actual objects in either the OOO sense or the Lacanian meaning: *they are the effects of objects*. In *Waste*, Brian Thill argues that "waste is every object, plus time" (New York: Bloomsbury, 2015, kindle loc. 120). And while Thill is not a Lacanian, he seems to affirm our thesis when he declares waste denotes

the fading of our desire: waste is "the unsatisfactory and temporary name we give to the affective relationships we have with our unwanted objects," it is "the expression of expended, transumuted, or suspended desire … the *ur*-object."

33. According to a recent symposium at the Museum of Modern Art (MoMA), there should be nine or even twelve recycling bins for plastics alone, as is done in Helsinki and Tokyo, respectively (Bruce Handy, "What's it Made of? Plastics," *New Yorker*, July 8 &15, 2019, 19).
34. That is, this algorithm (my formula; Fig. 5.2), which perhaps is a meta-algorithm (combining the semiotic rectangle and the formula of sexuation), argues that we cannot rely on algorithms, machine learning, to clean up the internet, to detect fake news, to scrub it of violence or hate speech—because that obscenity is constitutive of the internet, just as garbage is constitutive of nature or the environment.
35. In his commentary, Abalajon notes that the moderators' fallibility ought better to be laid at the door of highly mobile transnational companies (with precarious contracts that render unionization or a better-informed workforce unlikely) and a filmic style that settles for third world exoticism and contradictions ("slums vs high rise, religion vs consumerism") rather than the more mundane reality that these workers are just doing a job and trying to provide for their families. This would be a social reading not unlike the "exotic" Lacanian trope of *les non-dupes errent* (those who think they are not duped are in turn mistaken), which turns back on the film's audience, and this essay's author.
36. Fredric Jameson has suggested Lacan's "predilection for -isms and named thought systems stands as a fundamental example of what he termed *university discourse*," or the beginnings of "a theory of linguistic reification" in *Allegory and Ideology* (London: Verso, 2019), 99; emphasis in the original.
37. The narcissism of thinking "every bit counts" is the dialectical mirror-image of "I'm making it worse." Morton, for example, insists again and again that "every time I turned my car ignition key I was contributing to global warming" (see *Dark Ecology*, 8, 19, 35, 36 …). The hubris!
38. Here it is worth noting that before and during the time period of this book and the preceding conference (from 2013–2019), a minor international *contre-temps* took place between Canada and the Phillippines over a trans-Pacific shipment of garbage (103 containers) that had been sent from Vancouver to Subic Port, then sat in the harbour for years before

being returned to Canada. The containers were supposedly recycling but two-thirds of them turned out to contain ordinary household garbage. "Canadian garbage on its way from Philippines to Vancouver." https://www.cbc.ca/news/technology/canadian-garbage-from-philippines-departure-1.5156007 (accessed July 29, 2020).

Bibliography

Alain Badiou. 2018. *Lacan*. Translated by Kenneth Reinhard and Susan Spitzer. New York: Columbia University Press.

Barnard, Suzanne. 2002. Tongues of Angels: Feminine Structure and Other Jouissance. In *Reading Seminar XX: Lacan's Major Work on Love, Knowledge, and Feminine Sexuality*, ed. Suzanne Barnard and Bruce Fink, 171–185. Albany: State University of New York.

Burnham, Clint. 2018. *Does the Internet Have an Unconscious? Slavoj Žižek and Digital Culture*. New York: Bloomsbury.

Business for Social Responsibility (BSR). 2018. Human Rights Impact Assessment: Facebook in Myanmar, October. Accessed May 27, 2020. https://about.fb.com/wp-content/uploads/2018/11/bsr-facebook-myanmar-hria_final.pdf.

Canadian Garbage on Its Way from Philippines to Vancouver. Accessed July 29, 2020. https://www.cbc.ca/news/technology/canadian-garbage-from-philippines-departure-1.5156007.

Chiesa, Lorenzo. 2007. *Subjectivity and Otherness: A Philosophical Reading of Lacan*. Cambridge, MA: MIT Press.

Copjec, Joan. 1994. *Read My Desire: Lacan Against the Historicists*. Cambridge, MA: MIT Press.

Heffernan, Virginia. 2016. *Magic and Loss: The Internet as Art*. New York: Simon and Shuster.

Jameson, Fredric. 1969–1970. Walter Benjamin, or Nostalgia. *Salmagundi* 10–11: 52–68. https://www.jstor.org/stable/40546514.

———. 2019. *Allegory and Ideology*. London: Verso.

Kornbluh, Anna. 2016. States of Psychoanalysis: Formalization and the Space of the Political. *Theory & Event* 19 (3). Accessed July 19, 2019. https://www.muse.jhu.edu/article/623989.

Lacan, Jacques. 1998. *The Seminar of Jacques Lacan, Book XX: Encore, 1972–1973*. Edited by Jacques-Alain Miller. Translated by Bruce Fink. New York: Norton.

———. 2010. *The Seminar of Jacques Lacan, Book XX: Encore, 1972–1973.* Translated by Cormac Gallagher, from unedited French manuscripts. Eastbourne, UK: Antony Rowe. Accessed July 19, 2019. http://www.lacanireland.com/web/wp-content/uploads/2010/06/Book-20-Encore.pdf.
———. 2017. *The Seminar of Jacques Lacan, Book V: Formations of the Unconscious, 1957–1958.* Edited by Jacques-Alain Miller. Translated by Russell Grigg. Cambridge: Polity Press.
McGowan, Todd. 2004. *The End of Dissatisfaction? Jacques Lacan and the Emerging Society of Enjoyment.* Albany: State University of New York.
———. 2018. The Signification of the Phallus. In *Reading Lacan's Écrits: From Signification of the Phallus to Metaphor of the Subject*, ed. Stijn Vanheule, Derek Hook, and Calum Neill, 1–20. London: Routledge.
Morton, Timothy. 2016. *Dark Ecology: For a Logic of Future Coexistence.* New York: Columbia University Press.
Newton, Casey. 2019. The Trauma Floor: The Secret Lives of Facebook Moderators in America. *Verge*. February 25. Accessed July 22, 2019. https://www.theverge.com/2019/2/25/18229714/cognizant-facebook-content-moderator-interviews-trauma-working-conditions-arizona.
Roudinesco, Elisabeth. 1999. *Jacques Lacan: A Biography.* Translated by Barbara Bray. New York: Columbia University Press.
Žižek, Slavoj. 1992. *Enjoy Your Symptom! Jacques Lacan in Hollywood and Out.* New York: Routledge.
———. 1997. *The Plague of Fantasies.* London: Verso.

Part II

Our Knowledge on Climate Change

6

The Psychotopology of Climate

Sasha J. Langford

In "Anthropocene, Capitalocene, and the Problem of Culture," Daniel Hartley argues that Anthropocene discourse implies notions of time and space devoid of history, whereby a "naturalistic materialism …[belies] the reality of class struggle."[1] As Hartley suggests, one of the key sites in which the Anthropocene's ideological operations reproduce themselves is the realm of "scientific representations," which frame a materialist account of history as "practically unthinkable."[2] Indeed, as T. J. Demos discusses in *Against the Anthropocene: Visual Culture and Environment Today*, scientific images have "been central, even integral to the process of conceptualizing the Anthropocene," and such images, far from being neutral representations, carry with them ideological implications beyond their apparent legibility.[3] Ostensibly objective data visualisations, such as line graphs illustrating temperature increases, sea-level rise charts, and diagrams depicting the rate of melting arctic ice, rather than simply function to disseminate and defend evidence for anthropogenic global warming to

S. J. Langford (✉)
Department of Mass Communication, Columbia College,
Vancouver, BC, Canada

C. Burnham, P. Kingsbury (eds.), *Lacan and the Environment*, The Palgrave Lacan Series, https://doi.org/10.1007/978-3-030-67205-8_6

the public, obscure the capitalist relations that generate and sustain global warming, ultimately allowing the "military-state-corporate apparatus to disavow responsibility" for environmental devastation.[4]

While Demos turns to contemporary photography as an alternative visual modality to imagine a materialist account of ecology, for the purposes of this chapter, I consider the way in which the geometric—rather than solely aesthetic—principles of Anthropocene data visualisations contribute to their depoliticised and universalised concepts of history. I argue that the line graph—one of the most widespread scientific images associated with public awareness of climate change—serves to obscure the social relations at the core of the climate crisis in part due to the mapping of its data within the Euclidean space of the Cartesian plane, whereby relations can only be conceived through visual form rather than through intrinsic mathematical relationships. Against such a framework of time and space, I propose that topological approaches in Lacanian psychoanalysis offer an alternative spatio-temporal modality for thinking about climate change, which convey the significance of political relation in place of the generalised humanity of Anthropocene discourse.

The Line Chart and the Scientific Visualisation of Climate Change

Since Al Gore's documentary *An Inconvenient Truth* in 2006, the line graph has become one of the most widespread scientific images associated with the public awareness of climate change. While appearing in similar forms in the First Assessment Report published by the Intergovernmental Panel on Climate Change in 1990—a document which initiated the United Nations Framework Convention on Climate Change and the Kyoto Protocol—Gore's dramatic engagement with line charts prompted greater public awareness of the issue of climate change beyond the realms of policymakers and the scientific community.[5] Generally used to illustrate carbon dioxide levels, temperature anomaly, or ocean or surface temperature in relation to time, the line graph form figures as the dominant visual mode to represent data in a range of environmental infographics in news media, in non-specialist sources such as the popular science website *Welcome to the Anthropocene*, and as well as in authoritative documents such as each subsequent IPCC Assessment Report (Fig. 6.1).[6]

As Heather Houser argues in "The Aesthetics of Environmental Visualizations," line charts "have visually shaped current environmental debate" in a manner unlike other imagery used to provoke awareness of the climate crisis such as photography, artwork, and other varieties of data representation, because of manner in which "the line … intimates action and not just perspective" and "indicates a path back to" the initial causes of ecological collapse while at once depicting a projection of the future.[7] Yet despite visually rendering a progression of significant dates for the development of human-induced global warming, the line chart does little to disclose the historical events associated with said dates, nor the material, social, and economic forces responsible for them. While the graphs published in the IPCC reports consistently use 1850 as the starting date for the graphs—unlike Gore's graphs, which primarily frame the climate crisis either within the scope of the twentieth century or within a longer context of geological time[8]—little context besides passing references to "man's activities" and "pre-industrial times" situates the instigating factors of climate change.[9] While the graphs may, as Houser suggests,

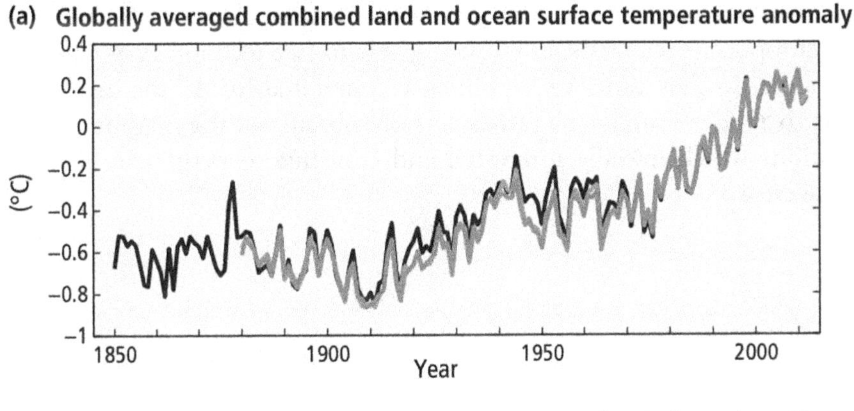

Fig. 6.1 IPCC, AR5 Synthesis Report 2014, Summary for Policymakers Figure SPM.1 [Detail], at https://www.ipcc.ch/report/ar5/syr/summary-for-policymakers/ (accessed October 1, 2019)

"[intimate] action" in the sense that they gesture towards a causal relation between the Industrial Revolution and rising greenhouse gas emissions, they do not provide information with regard to what or to whom such action may be attributed, nor to the social context in which this causal relation is taking place. With no other information given besides a discretely measured environmental effect and a span of dates, the line charts appear to assign causality of the climate crisis to a singular moment in history—a moment imagined outside of any particular geographic specificities, cultural dynamics, or social or economic modes of relation.

Visualising climate change as a one-to-one relation of empirical values rather than a historical process, the line chart paradoxically leaves little room to imagine the place of humans in relation to environmental collapse. Just as the concept of the Anthropocene distributes accountability for environmental devastation to a generalised notion of "humanity," while saying nothing about the social contexts that determine production and the use of resources, the line chart frames the climate crisis in a way that imagines a singular, inevitable relation between humanity, technology, and nature. Like Hartley and Demos' respective critiques of the Anthropocene as offering a universalised vision of history that denies the role of free-market capitalism in perpetuating the climate crisis, the visual form of the line graph facilitates the representation of a relation between time and the environment, or between time and technology, but never between humans in relation to themselves. Like Hartley's designation of "Anthropocene scientists" as not being able to "see technology as a political force, so they cannot see politics as a material force," the line chart fails to render visible—or rather actively obscures—the ongoing social relations which initially prompted and continue to perpetuate the climate crisis.[10]

Euclidean Space, Topography, and Universal History

While others such as Demos and Houser focus on the aesthetics of environmental data visualisations in the context of longer histories of scientific representation and discourses of objectivity, for the purposes of my discussion here I want to evaluate the role that geometric principles play in contributing to the line graph's ability to frame climate change in a way that evades an account of social and material dimensions. While including only a selective, singular quadrant from the full Cartesian coordinate system—including only that where the X and Y values are positive, and omitting the negative quadrants for the independent and dependent variables—the visual form of the line graph employs features of space first theorised by Euclid in the fourth century BC that still occupy a central place within Western thought. Developed into the Cartesian plane by Descartes in the seventeenth century in order to be able to define objects algebraically, Euclidean conceptions of time and space depend upon linear coordinates defined by their topographical relationships. As Virginia Blum and Anna Secor explain, "Euclidean geometry employs a metric understanding of space," and is "thus topographical; it refers to mappable, graphable, measurable space."[11] Unlike non-Euclidean geometries, where a potentially infinite number lines may pass through the same point, or where parallel lines are not employed at all, in the Euclidean space of the Cartesian coordinate system, only one line may pass through a given point where parallel lines will never intersect. As such, relations in Euclidean space depend upon visually quantifiable conceptions of space and time, where straight and curved lines are able to be drawn with a ruler and compass, and where one mathematical equation corresponds to only one possible visual representation of a geometric form.

Given the suitability of Euclidean space for visualising "mappable, graphable, measurable" space, it is perhaps not surprising that early developers of the line chart form as a mode of mapping data were concerned with making time and space into objects wholly definable within quantified terms. As discussed in a historical account of data visualisation, early developments of the line graph design took place amidst existing attempts in the seventeenth century to measure time and space, not just for the physical sciences and mathematics, but for "surveying, map making, navigation and territorial expansion."[12] The conception of space as coextensive with, and holding a one-to-one relation to its mathematical image is a necessary prerequisite for a conception of *place* as a tangible, material extension existing separate to the dynamics and histories of social life. As Maas and Morgan examine in a consideration of nineteenth-century graphical analysis, as well as communicating a conception of Cartesian "abstract space" within a broader European ideological context of imperial expansion, the modern form of the line graph reflects a cultural context in turn developing a notion of time as a quantified construction.[13] The "explosive growth" of the line graph form during the nineteenth century corresponds with a period of rapid industrialisation whereby the cyclical rhythms of agrarian life were increasingly replaced by rationalised management over the production process.[14] The popularisation of the line graph format required an understanding of time as an object ready for observation through discrete measurement, where events are "evaluated in standardised terms" rather than as part of a greater narrative of history.[15] The line graph makes sense in a social environment where history has been imagined "as data," where time as a subjective, concrete aspect of experience has been abstracted into numerical information.[16] Turning events into facts "which could be considered similar enough for valid comparison and which were measurable so that their numbers could be plotted in a Cartesian grid," the line graph both reflects and facilitates a worldview where time is inseparable from its quantification, where history can be understood from an indiscriminate and all-encompassing perspectival position.[17]

As sociologist Barbara Adam notes, a conception of time as abstract and quantified is in turn a necessary component for the temporality of capitalist production, as "all industrial time practices depend on time first being created to human design, that is, as abstract decontextualized and quantifiable clock-time" in order to be amenable to being "commodified, compressed and controlled."[18] In the case of the line graph, the form facilitates such a conception of time partly because of its early applications at the time of its modern emergence. The first economic line charts were published by Scottish engineer William Playfair within less than a decade of Adam Smith's *The Wealth of Nations*, the appearance of the first factory in Britain, and the invention of the steam engine by James Watt, the latter by whom Playfair was employed as a drafting technician during the same period.[19] As well as reflecting concerns for Britain's national deficits from military expenditures, Playfair's focus of his early charts on historical comparisons of national wealth and the price of wheat in Britain "spoke to the changing value and well-being of labor and to the balance between the agricultural and manufacturing sectors" whereby the management of time was increasingly instrumentalised for industrial production.[20] Despite initially only making use of proportional representations rather than numerical values, Playfair claimed that his graphs "could isolate and capture the essence of permanent causes" of history, proposing a visual correlate for the multifarious rhythms of lived experience within two-dimensional numerical space.[21] As a form that offers only one mathematically correct relation to be revealed by its data, Playfair's line charts present time as objectively available for division into a series of discrete components, rather than as contingent upon subjective perception, natural cycles, or cultural conditions. As it visualises "the 'progress' of abstract time as a dominant form of time," the line graph form in turn serves to communicate "the 'progress' of capitalism as a form of life," whereby the temporal dimensions of the future are both inevitable and singular.[22]

As with Houser's suggestion that the design choices of environmental information visualisations "not only have epistemological implications; they also make ethical appeals," in the context of climate change, the line graph derives part of its ideological potency from its topographical, Euclidean metric that does not readily subjectivise time, space, and environmental effects.[23] Rendering time as legibly distinct from history, the line graph is unable to account for the fluid inscriptions of lived experience, nor the divergent, multiple, and nonlinear nature of historical events. As such, the graph presents no distinction in the social distribution of its effects. The accelerated dangers of climate change for those living in countries lacking climate-resilient infrastructure and already prone to drought and malnutrition—in most instances because of legacies of colonial policies that have directly benefited Western nations—remain obscured by the graph's unified and continuous development of past, present, and future.[24] Proposing a visual correlate to imperialist narratives of "progress," the line graph uses a single time scale to generalise the effects of climate change, regardless of the specificity of its outcomes as they intersect with geography, race, class, gender, and other social factors. Using a visual language inherited from cartographic renderings and economic projections of the capitalist marketplace, the climate line chart presents the environmental disaster as formally identical to economic diagrams illustrating productivity and market boom, incidentally revealing the role of capitalism in propelling the forward movement of the climate crisis. Without reflexivity on this contradiction, the line chart constrains its projections within a universalised history and depoliticised humanity of the Anthropocene, where temporal experience has been naturalised within the quantified, abstract time of capitalist production.

Psychotopology: An Alternative Modality of Climate Change?

Against the topographical principles of the line graph, that, as I have suggested, communicate the generalised, depoliticised discourse of the Anthropocene of concern to Hartley and Demos, how might a different conception of space than that of Euclidean geometry contribute to alternative possibilities of understanding climate change? Against Hartley's assessment of dominant scientific visualisations, what would it take to make *thinkable* the material history of climate change within the field of temporal and spatial projections?

In contrast to the topographical space of Euclidean geometry and the Cartesian plane, I would like to consider the ways in which topological space, and particularly Jacques Lacan's theorisations of topology in psychoanalysis, permit a potential response to these questions. As Blum and Secor continue, topology defines geometric objects in terms of "sets of properties that retain their relationships under processes of transformation," whereby measurement is concerned with mathematical continuity rather than with consistency of visual metric.[25] Topological objects are studied for the unchanging or invariant relation of properties such as compactness, connectedness, and dimension, regardless of observable appearance.[26] As such, visual representations in topology studies are "[mistrusted]" or even "meaningless," as topological shapes generally defy the possibilities of concretely mappable modes of representation.[27] Unlike the direct use of the line and compass to understand and present space in Euclidean geometry, as topology concerns itself with "those properties of a thing that, while they are in a sense geometrical, are the most permanent" amidst distortion, a topological account of space relies upon intrinsic mathematical relation rather than graphical imagery.[28]

In the work of Lacan, the turn to topology has implications not just for realms of theoretical mathematics, but for understanding subjectivity, signification, and social relationships. Lacan employs topology's emphasis on invariant relation, rather than empirically observable manifestation, to account for the spatial and temporal dimensions of the unconscious that evade discrete measurement. For Lacan, mathematical structure constitutes the Real, and topology serves to conceptualise this structure by considering objects in space in terms of inherent properties rather than "homophonic resemblance."[29] It is by way of topology that Lacan can explain the deceptive role of visual continuity in the formation the subject, in which "identity and difference … are incessantly reflected in each other" and between which "there is no shared common space."[30] As it can positively formulate negative space, topology allows Lacan to assert the presence of *nonrelation*—sexual or otherwise—rather than simply relation's nonexistence. In other words, for Lacan topology visually renders the "support of the hole" within the subject; it formalises an approximation of the Real without serving a representational, figurative, or metaphorical function.[31]

For Lacan, the morphic transformation of space possible within topology accounts for the contradictory spatial dimensions of the unconscious, whereby conventional distinctions between interior and exterior cannot be distinguished, whereby the subject's "surface"-level slips of the tongue are constitutive of the depths of their inner life. Yet as well as these spatial qualities, for Lacan topological figures necessarily in turn also have temporal implications. The Möbius strip—a recurring topological figure invoked by Lacan—mimics the "temporal movement" of "the discourse of the unconscious," whereby in the confounding act of tracing one's finger along its surface, one inevitably ends up on its "flipside."[32] The unorientable surfaces of topological figures permit no progressive movements, but rather, like the dialectical temporality of the signifying chain, keep one "[going] round in circles."[33] Unlike the unilinear time of the line graph—which can only register quantifiable, positively-affirmed data—Lacan's use of topology as an explanatory tool offers the possibility to account for aspects of experience that evade verifiable appearance within conscious life. Just as any manifestation of the unconscious cannot be traced in a linear fashion back to a singular point of origin—but must rather be considered within a dynamic and recursive system whose parts are under continuous transformation—topology permits the retroactive temporality of the unconscious to be contemplated, it supports the seeming paradox of "the mechanisms of anticipation and retroaction" to be sustained without contradiction.[34]

As such, for Lacan topological diagrams such as the Möbius strip, the Klein bottle, and the Borromean knot are not aesthetic illustrations or metaphorical figures, they are rather formal structures that express the antagonisms of subjectivity at the ontological level. For Lacan, topology offers a framework to account for the way in which "the subject of the unconscious is evident but unacknowledged" within the conventional Cartesian *cogito*.[35] Contesting the assumption that conscious thought offers transparent and direct access to metaphysical truths with his assertion "where [the subject] is not, he is thinking," Lacan employs topology as a paradigm to summarise the *lack* of a singular, linear relation between ontology and epistemology that can be expressed by readily observable properties of space and time.[36] Topology rather renders conceivable a structure that accounts for the multiple localisation of the subject—a subject who, as "first and foremost the subject of language," remains caught in a contingent yet fixed dynamic between psychic and corporeal dimensions.[37]

In relation to climate change, the discontinuous temporality of Lacanian psychotopology offers a very different account of time than the line graph, one that may more authentically attest to the nonlinear dimensions of the environmental crisis. Against the progressive projections suggested by the line graph, as much scientific research has noted, the recursive dimensions of climate change, and the enormous amount of variables that contribute to its conditions, make future outcomes extremely difficult to predict with certainty.[38] Not only are the developments of the effects of climate difficult to linearly anticipate, but also, as eco-philosopher Timothy Morton argues, one of the unique characteristics of climate change is the way that at the experiential level it manifests itself in a contradictory manner that cannot be accounted for in a singular narrative of time or space:

> Global warming is not a function of our measuring devices. Yet because it's distributed across the biosphere and beyond, it's very hard to see as a unique entity. And yet, there it is, raining on us, burning down on us, quaking the Earth, spawning gigantic hurricanes. Global warming is an object of which many things are distributed pieces … it involves a massive, counterintuitive perspective shift to see it.[39]

As with Morton's own analogy elsewhere of "[talking] to the unconscious" as a mode of dealing with climate change, as this passage suggests, climate change, despite all the scientific data that exists about it, presents an empirical paradox when attempted to be approached directly.[40] Like Lacan's notion "where I am thinking … I am not," as Morton suggests, climate change *is* regardless of where it appears to be.[41] Climate change forms part of the consistency of contemporary experience, regardless of the presence or absence of any dramatic manifestations of it—tsunamis, hurricanes, and the like—that we might encounter in a given place and time. Just as attending to a single verbal expression, symptom, or mode of embodiment might be essential for trying to approach the unconscious—as psychoanalysts do when they listen to and observe their patients—yet not give by any means a full account of unconscious dynamics, it remains a contradiction that the immediacy of weather variance, while the only way to announce, encounter, and embrace the direct presence of climate effects concretely, remains a misrepresentation of climate's more volatile rhythms that remain beneath the threshold of awareness.

Precisely because, as Morton suggests, "global warming is not a function of our measuring devices," the line graph's modality of topographical space fails to account for its contradictory, nonlinear, and morphologically inconsistent dimensions which resist both empirical observation and visual representation. The universalised format of the graph does little to reveal the absential presence of climate change as described by Morton, which forms a mundane consistency of the everyday as much as it does tangible weather events. Like Lacan's critique of positivist science as having no room for negativity and thus encouraging the subject "to forget [himself]," the universalised format of the line graph makes it possible for climate deniers to use absential or nonlinear expressions of climate change to denounce its existence entirely.[42] Because it is visually and empirically elusive, because it ontologically resists the positivist logic of the *cogito*, climate change requires a formalisation that—like the subject of the unconscious—is defined by divergent and inconsistent temporalities that affirmatively mark negation.

Unlike the explanatory tool of the line graph format, topology proposes a mode of formalisation to understand climate change that does not aim to represent or summarise data, but is rather constitutive of a form of data itself, of the data which is itself form. As Anna Kornbluh discusses, Lacan's use of topology and mathemes can be considered part of the larger political possibilities of psychoanalysis, as these formalisations "[render] thinkable aspects of possible worlds which are not experienceable."[43] Reflecting upon, rather than simply being mediated by the space of the Symbolic, mathematical formalisations offer "a new kind of causality" that contain within themselves a metacommentary on their own operations.[44] In a context where popular understandings of climate change have been overdetermined by the language and images of institutional bodies like the IPCC, such political possibilities of formalisations like topology could respond to the urgency of making climate change thinkable in its full *unthinkability*; they may allow climate change to be approached not simply as an impending threat or a problem to be solved, but as a recursive, excessive, and indirectly-accessed component of lived reality that warps our very perception as we attempt to understand it.

The political possibilities of Lacan's use of topology further permits consideration of the role of relation inherent to the production of climate change, which have seen distortions in its visual representation over time. Against the consistent empirical manifestation suggested by the line graph, a social history of climate change suggests the phenomenon to have had many transformations in appearance, despite the fact that its intrinsic relations of exploitation have remained continuous. Among other scholars who have considered the inseparability of climate change and colonialism, Denise Ferreira da Silva most directly traces the material transformation of the warming of the atmosphere to a causal relation between white supremacy and capital:

> the accumulation of greenhouse gases in the troposphere corresponds to (as it is a transformation of) the extensive and intensive *extraction* of matter from the earth … [it] expresses (is equivalent to) the extent of expropriation and the intensity of the concentration of expropriated internal (kinetic) energy of lands and labour facilitated by coloniality and raciality.[45]

Unlike the Anthropocene's generalised attention to human technology use as the cause of the climate crisis, climate change, as Ferreira da Silva suggests, is rooted in the material effects of a structure of social relations, those of "coloniality and raciality." Just as coloniality has historically depended upon racialised notions of climate, anthropogenic climate has depended upon and been cocreated with coloniality. There is a precise correspondence with the exploitation of racialised bodies in the form of slave labour, and the production of environmental outcomes. The current experience of the climate crisis is materially rooted in "colonial juridic mechanisms and racial symbolic tools" as systems of discourse, which have turned the energetic matter of "the internal energy of African slaves" into the material effect of heat within the atmosphere.[46] Climate change is not a technological problem, nor a humanist one, but the historical result of relations of racialised exploitation and displacement fundamental to the capitalist state, that have had shifted appearances throughout their history.

Lacanian psychotopology, in its insistence on relation—a relation that as a *non-relation*, is, as Samo Tomšič suggests, the opposite of "the capitalist social relation founded on abstract freedom and equality"—permits a mode of thought that can accommodate and render conceivable this morphological transformation of climate and capital of which Ferreira da Silva speaks.[47] Unlike the attempts of Anthropocene discourse to propose technological solutions, "sustainability," or even "green" capitalism to combat the climate crisis, a topological approach to climate change helps expose that such responses are merely visual distortions of the same relations of exploitation. Like Marx's own desire to "[uncover] the immanent breaks in existing social appearances" and displace the inversion of essence and appearance invoked by the commodity form, topology offers a modality that can overturn the epistemological assumptions of dominant climate change discourse that maintain a depoliticised vision of humanity. By formally conceiving the dimensions of space and time as relation rather than as empirically observable appearance, topology offers a modality for climate change that makes thinkable the contradictions of history in their full materiality.[48]

Against Demos' proposition of a turn to contemporary photographic practices to socialise the visual discourse of climate change, topology, and Lacan's employment thereof, permits a modality for considering climate change that does not depend upon aesthetic dimensions, but rather on inherent relation. As such, it problematises the ideology of Anthropocene discourse and the line graph, which conceive of the ecological crisis solely through the measured presence of environmental effects. Able to give positive presence to absence, and readily accommodate the nonlinear temporality of climate change, a psychotopological account of climate may help to encourage a politics that moves away from the visible in order to reveal climate as not a merely a matter of images, but of justice.

Notes

1. Daniel Hartley, "Anthropocene, Capitalocene, and the Problem of Culture," in *Anthropocene or Capitalocene?: Nature, History, and the Crisis of Capitalism*, ed. Jason W. Moore (Oakland, CA: PM Press, 2016), 155.
2. Hartley, "Anthropocene, Capitalocene," 157.
3. T. J. Demos, *Against the Anthropocene: Visual Culture and Environment Today* (Berlin, Sternberg Press, 2017), 15.
4. Demos, *Against the Anthropocene*, 17.

5. Heather Houser, "The Aesthetics of Environmental Visualizations: More Than Information Ecstasy?" *Public Culture* 26, no. 2 (2014): 319–37, https://doi.org/10.1215/08992363-2392084. See also Al Gore, *An Inconvenient Truth*, directed by Davis Guggenheim (Los Angeles: Paramount Classics, 2006).
6. Nicholas W. Smith and Hélène Joffe, "Climate Change in the British Press: The Role of the Visual." *Journal of Risk Research* 12, no. 5 (2009): 647–63. https://doi.org/10.1080/13669870802586512.
7. Houser, "Aesthetics of Environmental Visualizations," 327–28.
8. As Houser notes, the scale of latter being particularly well-suited to literally "dramatizing the expression 'off the charts'" (Houser, "Aesthetics of Environmental Visualizations," 322).
9. World Meteorological Organization/United Nations Environment Programme, *Climate Change: The IPCC 1990 and 1992 Assessments*, Intergovernmental Panel on Climate Change. June 1992, 79, 52, https://www.ipcc.ch/site/assets/uploads/2018/05/ipcc_90_92_assessments_far_full_report.pdf (accessed October 1, 2019).
10. Hartley, "Anthropocene, Capitalocene," 157.
11. Virginia Blum and Anna Secor, "Psychotopologies: Closing the Circuit between Psychic and Material Space." *Environment and Planning D: Society and Space* 29, no. 6 (2011): 1034, https://doi.org/10.1068/d11910.
12. Michael Friendly and Daniel J. Denis, "Milestones in the History of Thematic Cartography, Statistical Graphics, and Data Visualisation," November 2006, 7, http://datavis.ca/milestones/ (accessed May 27, 2020)
13. Sarah A. Radcliffe, "Re-Mapping the Nation: Cartography, Geographical Knowledge and Ecuadorean Multiculturalism," *Journal of Latin American Studies* 42, no. 2 (2010): 306, https://www.jstor.org/stable/40784984.
14. Friendly and Denis, "Milestones," 15.
15. Harro Maas and Mary S. Morgan, "Timing History: The Introduction of Graphical Analysis in 19th Century British Economics," *Revue D'Histoire Des Sciences Humaines* 7, no. 2 (2002): 97, https://www.cairn.info/revue-histoire-des-sciences-humaines-2002-2-page-97.htm (accessed, October 1, 2019).
16. Maas and Morgan, "Timing History," 97.
17. Maas and Morgan, "Timing History," 97.
18. Barbara Adam, *Time* (Cambridge: Polity Press, 2004), 73.
19. Maas and Morgan, "Timing History," 97.

20. Judy L. Klein, "Reflections from the Age of Economic Measurement," *History of Political Economy* 33 Annual Supplement (2001): 112, https://muse.jhu.edu/article/13335/pdf.
21. Klein, "Reflections," 112.
22. Moishe Postone, quoted in Jonathan Martineau, *Time, Capitalism and Alienation: A Socio-Historical Inquiry Into the Making of Modern Time* (Chicago: Haymarket Books, 2016), 107.
23. Houser, "Aesthetics of Environmental Visualizations," 321.
24. Suzanne Goldenberg, "Climate Change: The Poor Will Suffer Most," *Guardian*. March 31, 2014, https://www.theguardian.com/environment/2014/mar/31/climate-change-poor-suffer-most-un-report (accessed January 9, 2020); and United Nations General Assembly, "Unprecedented Impacts of Climate Change Disproportionately Burdening Developing Countries, Delegate Stresses, as Second Committee Concludes General Debate," GA/EF/3516, 74th Session, October 8, 2019, https://www.un.org/press/en/2019/gaef3516.doc.htm (accessed January 9, 2020).
25. Blum and Secor, "Psychotopologies," 1034.
26. James R. Munkres, *Topology* (New York: Pearson, 2018).
27. Stephen Barr, *Experiments in Topology* (New York: Dover, 1989), 2.
28. Barr, *Experiments in Topology*, 2–3.
29. Rona Cohen, "The Spatiality of Being: Topology as Ontology in Lacan's Thinking of the Body," in *Psychoanalysis: Topological Perspectives: New Conceptions of Geometry and Space in Freud and Lacan*, ed. Michael Friedman and Samo Tomšič (Bielefeld, Germany: transcript-Verlag, 2016), 240.
30. Slavoj Žižek, *Sex and the Failed Absolute* (London: Bloomsbury Academic, 2020), 223–24.
31. Jacques Lacan, *Seminar XXIII: Joyce and the Sinthome Part 1; Joyce and the Sinthome Part 2*, trans. Cormac Gallagher, 5, http://www.lacaninireland.com/web/wp-content/uploads/2010/06/THE-SEMINAR-OF-JACQUES-LACAN-XXIII.pdf (accessed January 9, 2020).
32. Mai Wegener, "Psychoanalysis and Topology: Four Vignettes," in *Psychoanalysis: Topological Perspectives: New Conceptions of Geometry and Space in Freud and Lacan*, ed. Michael Friedman and Samo Tomšič (Bielefeld, Germany: transcript-Verlag, 2016), 35.
33. Jacques Lacan, *L'insu que sait…Seminar XXIV Final Sessions 1–12 1976–1977*, trans. Cormac Gallagher, 12, http://www.lacaninireland.

com/web/wp-content/uploads/2010/06/insu-Seminar-XXIV-Final-Sessions-1-12-1976-1977.pdf (accessed January 9, 2020).
34. Hoens, "Why Topology?" 58.
35. Calum Neill, *Lacanian Ethics and the Assumption of Subjectivity* (New York: Palgrave Macmillan, 2011), 17.
36. Jacques Lacan, *The Seminar of Jacques Lacan, Book XVII: The Other Side of Psychoanalysis*, ed. Jacques-Alain Miller, trans. Russell Grigg (New York: W. W. Norton, 2008), 103.
37. Neill, *Lacanian Ethics*, 17.
38. For example, see Nature Climate Change, "Certain Uncertainty" (editorial), *Nature Climate Change* 3, no. 689 (2013), https://doi.org/10.1038/nclimate1978; and Clara Deser, Adam Phillips, Vincent Bourdette, and Haiyan Teng, "Uncertainty in Climate Change Projections: The Role of Internal Variability," *Climate Dynamics* 38, nos. 3–4 (2012): 527–46, https://doi.org/10.1007/s00382-010-0977-x; among others.
39. Timothy Morton, *Hyperobjects: Philosophy and Ecology after the End of the World*. Minneapolis: University of Minnesota Press, 2014, 44.
40. Morton, *Hyperobjects*, 184.
41. Lacan, *Book XVII: The Other Side of Psychoanalysis*, ed. Miller, trans. Grigg, 103.
42. Jacques Lacan, "The Function and Field of Speech and Language in Psychoanalysis," in *Écrits: A Selection*, ed. Jacques-Alain Miller, trans. Bruce Fink (New York: W. W. Norton, 2006), 233.
43. Anna Kornbluh, "States of Psychoanalysis: Formalization and the Space of the Political," *Theory & Event* 19, no. 3 (2016), https://www.muse.jhu.edu/article/623989.
44. Kornbluh, "States of Psychoanalysis."
45. Denise Ferreira da Silva, "On Heat," *Canadian Art* (online edition), October 29, 2018, emphasis in the original; https://canadianart.ca/features/on-heat/ (accessed October 1, 2019).
46. Ferreira da Silva, "On Heat."
47. Samo Tomšič, *The Capitalist Unconscious: Marx and Lacan* (London: Verso, 2015), 28.
48. Tomšič, *Capitalist Unconscious*, 29.

Bibliography

Adam, Barbara. 2004. *Time*. Cambridge: Polity Press.

Barr, Stephen. 1989. *Experiments in Topology*. New York: Dover.

Blum, Virginia, and Anna Secor. 2011. Psychotopologies: Closing the Circuit between Psychic and Material Space. *Environment and Planning D: Society and Space* 29 (6): 1030–1047. https://doi.org/10.1068/d11910.

Cohen, Rona. 2016. The Spatiality of Being: Topology as Ontology in Lacan's Thinking of the Body. In *Psychoanalysis: Topological Perspectives: New Conceptions of Geometry and Space in Freud and Lacan*, ed. Michael Friedman and Samo Tomšič, 227–250. Bielefeld, Germany: transcript-Verlag.

Demos, T.J. 2017. *Against the Anthropocene: Visual Culture and Environment Today*. Berlin: Sternberg Press.

Deser, Clara, Adam Phillips, Vincent Bourdette, and Haiyan Teng. 2012. Uncertainty in Climate Change Projections: The Role of Internal Variability. *Climate Dynamics* 38 (3–4): 527–546. https://doi.org/10.1007/s00382-010-0977-x.

Ferreira da Silva, Denise. On Heat. *Canadian Art* (online edition). October 29, 2018. Accessed October 1, 2019. https://canadianart.ca/features/on-heat/.

Friendly, Michael, and Daniel J. Denis. 2006.. Milestones in the History of Thematic Cartography, Statistical Graphics, and Data Visualisation. November. Accessed May 27, 2020. http://datavis.ca/milestones/.

Goldenberg, Suzanne. Climate Change: The Poor Will Suffer Most. *Guardian*. March 31, 2014. Accessed January 9, 2020. https://www.theguardian.com/environment/2014/mar/31/climate-change-poor-suffer-most-un-report.

Gore, Al. 2006. *An Inconvenient Truth*. Directed by Davis Guggenheim. Los Angeles: Paramount Classics.

Hartley, Daniel. 2016. Anthropocene, Capitalocene, and the Problem of Culture. In *Anthropocene or Capitalocene?: Nature, History, and the Crisis of Capitalism*, ed. Jason W. Moore, 154–165. Oakland, CA: PM Press.

Hoens, Dominiek. 2016. Why Topology? In *Psychoanalysis: Topological Perspectives: New Conceptions of Geometry and Space in Freud and Lacan*, ed. Michael Friedman and Samo Tomšič, 53–62. Bielefeld, Germany: transcript-Verlang.

Houser, Heather. 2014. The Aesthetics of Environmental Visualizations: More Than Information Ecstasy? *Public Culture* 26 (2): 319–337. https://doi.org/10.1215/08992363-2392084.

Klein, Judy L. 2001. Reflections from the Age of Economic Measurement. *History of Political Economy* 33 (Annual Suppl.): 111–136. https://muse.jhu.edu/article/13335/pdf.

Kornbluh, Anna. 2016. States of Psychoanalysis: Formalization and the Space of the Political. *Theory & Event* 19 (3) https://www.muse.jhu.edu/article/623989.

Lacan, Jacques. 2006. The Function and Field of Speech and Language in Psychoanalysis. In *Écrits: A Selection*, ed. Jacques-Alain Miller and translated by Bruce Fink, 197–268. New York: W. W. Norton.

———. *L'insu que sait...Seminar XXIV Final Sessions 1–12 1976–1977*. Translated by Cormac Gallagher. Accessed January 9, 2020. http://www.lacaninireland.com/web/wp-content/uploads/2010/06/insu-Seminar-XXIV-Final-Sessions-1-12-1976-1977.pdf.

———. 2008. *The Seminar of Jacques Lacan, Book XVII: The Other Side of Psychoanalysis*. Edited by Jacques-Alain Miller. Translated by Russell Grigg. New York: W. W. Norton.

———. *Seminar XXIII: Joyce and the Sinthome Part 1; Joyce and the Sinthome Part 2*. Translated by Cormac Gallagher. Accessed January 9, 2020. http://www.lacaninireland.com/web/wp-content/uploads/2010/06/THE-SEMINAR-OF-JACQUES-LACAN-XXIII.pdf.

Maas, Harro, and Mary S. Morgan. 2002. Timing History: The Introduction of Graphical Analysis in 19th Century British Economics. *Revue D'Histoire Des Sciences Humaines* 7 (2): 97–127. Accessed October 1, 2019. https://www.cairn.info/revue-histoire-des-sciences-humaines-2002-2-page-97.htm.

Martineau, Jonathan. 2016. *Time, Capitalism and Alienation: A Socio-Historical Inquiry Into the Making of Modern Time*. Chicago: Haymarket Books.

Morton, Timothy. 2014. *Hyperobjects: Philosophy and Ecology after the End of the World*. Minneapolis: University of Minnesota Press.

Munkres, James R. 2018. *Topology*. New York: Pearson.

Nature Climate Change. 2013. Certain Uncertainty (editorial). *Nature Climate Change* 3 (689). https://doi.org/10.1038/nclimate1978.

Neill, Calum. 2011. *Lacanian Ethics and the Assumption of Subjectivity*. New York: Palgrave Macmillan.

Radcliffe, Sarah A. 2010. Re-Mapping the Nation: Cartography, Geographical Knowledge and Ecuadorean Multiculturalism. *Journal of Latin American Studies* 42 (2): 293–323. https://www.jstor.org/stable/40784984.

Smith, Nicholas W., and Hélène Joffe. 2009. Climate Change in the British Press: The Role of the Visual. *Journal of Risk Research* 12 (5): 647–663. https://doi.org/10.1080/13669870802586512.
Tomšič, Samo. 2015. *The Capitalist Unconscious: Marx and Lacan*. London: Verso.
United Nations General Assembly. Unprecedented Impacts of Climate Change Disproportionately Burdening Developing Countries, Delegate Stresses, as Second Committee Concludes General Debate. GA/EF/3516. 74th Session. October 8, 2019. Accessed January 9, 2020. https://www.un.org/press/en/2019/gaef3516.doc.htm.
Wegener, Mai. 2016. Psychoanalysis and Topology: Four Vignettes. In *Psychoanalysis: Topological Perspectives: New Conceptions of Geometry and Space in Freud and Lacan*, ed. Michael Friedman and Samo Tomšič, 31–52. Bielefeld, Germany: transcript-Verlag.
World Meteorological Organization/United Nations Environment Programme. 1992. *Climate Change: The IPCC 1990 and 1992 Assessments*. Intergovernmental Panel on Climate Change. June. Accessed October 1, 2019. https://www.ipcc.ch/site/assets/uploads/2018/05/ipcc_90_92_assessments_far_full_report.pdf.
Žižek, Slavoj. 2020. *Sex and the Failed Absolute*. London: Bloomsbury Academic.

7

In Defence of the Subject

Juan Luis de la Mora

Tlalticpac toquichtin tiez
(The earth shall be as men are)
—Náhuatl proverb

This text assumes that global environmental chaos is a consequence of capitalism, of its radical incompatibility with life. I will try to show some of the subjective consequences of the strategies that capitalism deploys to ensure its expansion, and how these consequences relate to the seemingly unstoppable destruction of life and its diversity. As consumption has become the practice that institutes subjectivity, capitalism has developed a series of micropolitical operations to secure the conditions for its reproduction.[1] In order to ensure continuous consumption, capitalism captures and abuses the potency of desire, resulting in what I call subjective depredation: the attrition of the subject's symbolic and imaginary resources. To show how these processes relate to environmental chaos,

J. L. de la Mora (✉)
Mexico City, Mexico

C. Burnham, P. Kingsbury (eds.), *Lacan and the Environment*, The Palgrave Lacan Series, https://doi.org/10.1007/978-3-030-67205-8_7

I will start by discussing how Modernity configured nature as a deposit of resources, to be exploited for the benefit not of humankind but of capital; this will allow us to think about the production of subjectivities under the discourse of the markets. I will then discuss subjective depredation, some of its consequences, and how the markets operate to reproduce it. Finally, I will present an example of artistic production to discuss how the recovery of the subject's resources can contribute to the possibilities of revolt against capitalism.

Nature appears to us as something that has always been there, identical to itself, impervious to the way we approach it. Modern science has taught us to conceive the physical world as a collection of objects and as a network of phenomena connected by the laws of cause and effect. We tend to ignore that nature—as much as subjectivity, territory, or time—is a product of discourse. Our nature is a realisation of Western *ontotheo-politics*, more specifically the civilising project of Modernity.

In a 2007 text on biopolitics, Juan Camilo Cajigas-Rotundo shows how the colonial matrix of power not only produces a certain conception of nature, but also depends on it to reproduce itself. He traces this construction of nature as far back as Descartes.[2] Albeit saying that his pen gave birth to the modern philosophical subject might be an exaggeration, Descartes' *cogito* reveals this self-sufficient subject as a curious observer. Reason provided a stable footing to explore and dissect a deceiving, fickle, ever-changing world. The subsequent isolation of this subject as a reasoning individual is the origin of what Santiago Castro-Gómez calls zero-point hubris.[3] This author proposes a subject that looks upon the world like a god: He can see everything, and he cannot be seen; he is faceless, without attributes. This vantage-point endows him with an aura of universality, a camouflage—even to himself—of his historical conditions. European Enlightenment and Newtonian physics, modern scientific thought, allowed us to set (as one sets a trap) mathematical laws that can describe the inner workings of nature. Cartesian dualism allowed for Modernity's program of human domination of nature. This historical unfolding of modern Western thought is indivisible from the planetary development of capitalism. I think of capitalism not only as an economic system of production and distribution of wealth, but also as a way of producing life itself. In Karl Marx's words:

> The mode of production of material life conditions the general process of social, political and intellectual life. It is not the consciousness of men that determines their existence, but their social existence that determines their consciousness.[4]

As the mode of production of material life developed into contemporary capitalism, not only social relations become dominated by the "natural laws" of the market, as Marx himself calls them. Nature itself changed; it went from divine creation—a gift from god to humankind—to a network of measurable correlations, so its telos went from divine will to the action of humans.[5]

Martin Heidegger studied a particular facet of this shift in Western history in *The Question Concerning Technology*.[6] In traditional artisanal technology, nature is taken care of so that it reveals what it conceals. Heidegger calls this pro-duction: to-take-forwards, to conduct something to its place. An artisan farmer takes care of the soil and the seed, but he can only abandon himself to the processes of germination and growth. With industrial agriculture, nature is no longer left to its own rhythms, it is no longer pro-duced but *pro-voked* or challenged; it is now forced unreasonably, in Heidegger's own terms, to yield the energy it contains so that it can be harvested, exploited, and accumulated. This mode of unconcealment closes the possibility of creation, art, and, ultimately, transforming *poiesis.*

Imperial colonialism set the material foundations for the development of capitalism. As Western Modernity covered the planet, it configured nature as a trove of exploitable resources, devouring all other cosmopolitical conceptions along the way. Indigenous peoples are under siege as they try to defend not only their traditional ways of life, but also non-Westernised ways of relations with non-human life and, more generally, the cosmopolitical fabric of their worlds. Global Witness reported that there were 172 murdered environmental defenders in 2018, and that countless others are persecuted and criminalised.[7] In a broader sense, these killings are part of the necropolitics at the centre of the continuous deployment of the colonial matrix of power. The universalism of Western thought translates to the engulfing, all-devouring movement of colonialism. As Modernity unifies the world, it neutralises or eliminates cultural diversity whenever it hinders the expansion of capital. There is a connection between the systemic destruction of biodiversity and this fear of cosmopolitical diversity; a correlation between the constitution of the modern subject, fiercely individualistic in its negation of historical conditions, the subsequent dynamic of voracious and violent unification under the flag of development, and our unreasonable and stubborn failure to value and protect biodiversity.

When we listen to indigenous peoples and environmental activists, we can identify three forms of depredation that contribute to environmental chaos: environmental depredation, social depredation, and depredation of labour.[8] I will describe them very briefly, for I can only rely on what I have understood by listening to their stories. Then, I will propose a fourth kind of depredation, where I think the micropolitical potency of psychoanalytical theory and practice may prove of some value.

Environmental depredation consists of the impoverishment or destruction of the conditions necessary to sustain life. Loss of habitats is as menacing to biodiversity as climate change. Several practices contribute to environmental depredation, such as extractive exploitation activities; industrial farming reliant on monocropping, pesticides, and deforestation; water and air pollution, and so on.

Social depredation is closely correlated, as it is necessary to dissolve social links in order to displace or subdue populations that might stand in the way of so-called *development*. Disarticulation of traditional practices and knowledge is also an active policy of modern states, as they need to construct a cohesive collective identity. Diversity of languages hinders state governance, and native speakers can be as dangerous as human rights activists and environmental defenders.[9] Epistemic extractivism is another form of social depredation: governments or private entities appropriate traditional practices and knowledge to commodify them. Examples include textiles, which are often textual or ritual expressions of culture; and, most notably, plants with medicinal and psychedelic properties that have been patented by pharmaceutical laboratories.[10] These practices result in the weakening of ties of solidarity, identity, and community.

Depredation of labour is also called precarisation of employment, although it is life itself that is being precarised, since labour is the form in which the conditions for social life are reproduced under capitalism. Labour rights are violently reduced in the name of *austerity*, mechanisation is intensified by the introduction of pro-gramming, and evermore complex systems of interconnection that obey impersonal masters. Depredation of labour takes different forms in different latitudes, as international division of labour and distribution of wealth continue to have such effects, but the result is always the same: life is cornered, its potency stifled.

There is a fourth form of diminishment of life's diversity: subjective depredation. One effect of this form of depredation was detected by Lacan in 1953, when he denounced "anhistorism" as a cultural characteristic of the "American way of life," radically opposed to the experience of psychoanalysis. This negation of the possibility of communal and subjective historisation, is at the core of "behavioural human engineering."[11] More than 15 years later, when he developed his theory of the four discourses, Lacan further explored the consequences of capitalism on subjectivity. In February 1970, he told his audience that imperialistic colonialism is a form of the discourse of the master, as he referred to three of his analysands, Togolese men who had received formal education in France. They had reconfigured their childhood memories retroactively according to Western Oedipal categories, ultimately losing connection with their culture and traditions, which they knew solely through French colonial ethnography.[12]

But how does this resistance to historisation works? Why would the reproduction of capital depend on it? And what are the micropolitical strategies that allow for it to happen? Since the 1970s, globalised markets—a late realisation of what Lacan dubbed the capitalist discourse—have become the central referent of the process of subjective institution, a place held until then by the modern sovereign state, with its principles of development, destiny, identity, and continuity. Michel Foucault described confinement *dispositifs* of subjectivation as the series that takes each individual from family to school, from school to university, factory, office, or garrison; and, in cases of deviation, to the hospital or prison.[13] The state coordinated these instances, guaranteeing that a subject that underwent a particular process of education or normalisation in one space of discipline would be delivered to the next with certain expected characteristics and competencies.

Things change when the acephalous power of the markets determines the production of subjectivity. It is in what Gilles Deleuze characterised as societies of control—a term he took from William Burroughs—that subjective depredation takes its full scope.[14] In disciplinary societies, the primary support of subjectivity was the citizen, whereas in societies of control this function falls onto the consumer.[15] For Ignacio Lewkowicz, an Argentinian historian of subjectivity, "consumer" does not point to an accidental or local characteristic of a transhistorical human nature, but rather defines a concise and historically localised result of the practices that institute subjectivity.[16] He takes a Foucaultian approach to the analysis of social symptoms, by studying historical mutations of social practices in relation to instituted subjectivity. In hypermodern capitalism, it is the practice of consumption that produces and localises a subject, in the same way that nationality, conscience, and freedom were the coordinates that defined modern subjectivities. However, instances of consumption do not need to be connected as disciplinary spaces were, and the images that an individual constructs in each instance do not necessarily form a coherent identity. In the seemingly infinite series of objects that are offered to a consumer, the new is valuable because it is new, and the object that loses its value does not establish any historical relationship with the consumer. Time is also a construction determined by each historical situation, and the time produced by consumption is the time of fashion. In fashion, what counts is the apparently inexhaustible supply of objects to consume. Each element opens the infinity of the entire series: the market is fabulous because it operates under the assumption that everything and anything can be found. The market appears without lack. Each object, when it occupies the place of what-is-fashionable, promises an ecstatic state of imaginary completeness. The logic of consumption is based on the promise of a complementarity between subject and object that would bypass the division of the subject by language, the irreparable loss that produces object *a* as residual cause of desire. When someone browses through Amazon, Spotify, or Tinder, they have the sensation that there is something for everyone, for every situation, for every need. Advertising is the promise of enjoyment without limits. However, as this can never happen—because it is impossible and because compulsory consumption depends on it—there is no option but to look for the newest element of the series, sustained by the same imaginary logic. Consequently, time is no longer the continuous, transcendent, and progressive time of Modernity; instead, the time of fashion is the time of the instant; it is vertiginous, fragmentary, and discontinuous.

Those who inhabit this disjointed time find an enormous resistance to historisation. Fashion demands the subject to abandon the object of their supposed desire at every turn, without ever establishing any relationship with or between them; there is no sequence, only substitution. However, the logic of consumption, with its promise of completeness, prevents us from recognising our dissatisfaction. Being bored, sad, or in a bad mood is treated as a sign of stupidity: One needs simply to reach out and pick up a phone, play a game, or check a social media feed; or, even better, take a pill—whatever one needs to feel distracted or laugh or at least fall asleep. At the slightest itching of dissatisfaction we run away towards one of those captivating contraptions, mirrors, or trinkets that offer to restore an unattainable state of bliss. Instead, what is being silently established is a radical disconnection with everything and everyone—our discomfort *provokes* us to take out our phone and, without knowing what we are looking for, we check WhatsApp, Facebook, or Tinder. While trying to find connection we turn our back to ourselves and to the presence of others around us. Pain, in the absence of language—because who needs language when you can swipe right—cannot become suffering.[17] These are some of the ways in which the capitalist discourse diminishes the subject's capacity to inhabit experiences and to support a painful lack that produces desire. I will now try to further explore the importance of these micropolitical consequences in relation to environmental chaos.

Consumption captures desire. We can see this as algorithms track our every move to predict our interests (not only shopping, but love, sex, and even political preferences are being studied by algorithms). Is this not Lacan's behavioural engineer at its best? In his seminar on "The Purloined Letter," Lacan studies automatism and the determination of subjectivity by language. To explain repetition he described an autonomous machine made of signifiers, built upon the alternate function of absence and presence. Lacan asked his students to play a version of "even and odd," trying to guess each other's choices. One of them, knowing that randomness is impossible to imitate, followed the syllables of a poem by Mallarmé to choose at each turn, intending to bypass his unconscious subjective determinations. He ceded his power of decision to the poem. Lacan, surprised by his student's ruse, imagined a "modern calculating machine" that could decipher the determinative logic behind a subject's desires:

> If the unconscious exists in the sense of Freud,… if we listen to … the lessons he obtained from the experiences of the psychopathology of everyday life,… it is not unthinkable that a modern calculating machine, by deciphering the phrase that, unbeknownst to him, modulates … the choices of a subject, could win beyond any usual proportion in the game of odd and even.[18]

It is not relevant if these algorithms can predict consumer behaviour, because they can do something better: They can determine it. Desire, as Lacan pointed out numerous times, is desire of the Other, and our reticular *dispositifs* can function as the Other that *provokes* desire, robbing us of our singular and collective potency.[19] Depredation of subjective resources is an essential gear in the infernal, self-devouring machine that is capitalism. It is the violent imposition of *nuda vida*, not only in the streets of Ciudad Juárez or Gaza, in Ceuta and Melilla, in Aleppo or Mogadishu, but in Detroit and Igloolik as well, and even in London and Washington, in the bodies of the depressed and the addicted, in the hearts of thriving modern cities.

* * *

Fyodor Dostoevsky warned us about the power of what Sigmund Freud would later call *Trieb*, the drive. Its blind insistence can overpower reason, knowledge, and will. In its most enigmatic manifestations, it seems as if its survival was more important to the subject than that of the subject itself. That is the death drive: The prevalence of satisfaction over life. We can detect its effects in our clinical practice, in the weakening of the social ties of identity and solidarity, and in the way we continue to forcibly dominate nature.

In *Notes from the Underground*, Dostoevsky's nihilist, miserable narrator, directs his diatribe to an audience of technocrats, whom he admires and despises. At the beginning of Chap. 7, he ridicules the premise of the statistical calculation and classification of advantages as the basis for a behavioural prescription (recall Lacan's engineering), a scientifically determined moral code of conduct (for example, an algorithm). He spouts his bile insistently: How can it be that, more often than not, humans knowingly rush into a path of danger and lunacy, instead of the one marked as advantageous and safe? Maybe darkness and self-destruction are more *pleasurable* than advantage and self-preservation. He vows to follow the path of pleasure even if it entails annihilation—if only to upset those responsible for this new possibility of moral calculation.

> Why do all these statisticians, sages and lovers of humanity, when they reckon up human advantages invariably leave out one? … But the trouble is, that this strange advantage does not fall under any classification and is not in place in any list.[20]

This advantage, this factor that cannot be classified or inscribed, what can it be if not the real of *jouissance*? The amoral, stubborn, vibrant satisfaction of the drive? Capitalism is not only about the production and consumption of commodities and the accumulation of wealth. Economy is also economy of life and death, of desire and drive. There is no capitalism without enjoyment. This word is strongly related with consumption. In French, *jouir* refers almost immediately to sexual pleasure; while the English *enjoyment* is closer to happiness. I point this out because it can help us consider how capitalism seizes our bodies, our affections, and social spaces. Taking from Baruch Spinoza's *Ethics*, Deleuze talked about sad affections, those that arise from a sense of impotence. Capitalism needs those affections because they help reproduce consumption. "Power opposes potency."[21] If we feel sad and inadequate, we are more susceptible to believe that a particular commodity embodies the object of our lack, as if it were a missing piece of a puzzle. This is what Andy Fisher calls "necessary dissatisfaction."[22]

The endless beat of consumption is rooted in certain ways of producing and administering life and death—and of relating to nature. In societies where the capitalist mode of production prevails, production and consumption of commodities and ways of life are explained by the realisation of wealth as much as by *jouissance*. Only by taking this into account can we understand the cynicism of our individual and collective responses in the face of disaster. We know what is coming, but we do not want to stop. Cynicism is a product of the encounter of this knowledge and capitalism, insofar as it reproduces bad conscience and fake, empty enthusiasm. Cynicism is the other side of fear; it is a sad affection. We have enough information to know the scope of the myriad problems that environmental chaos will bestow upon life. We have enough material resources to start mitigation and adaptation, to conserve life's beauty and diversity. But we prefer to devise convoluted detours whenever necessary to maintain the practices that allow for the extraction and accumulation of wealth.[23]

* * *

It is safe to say, then, with Leonard Cohen, that the war is over and the good guys lost. It may be poetic license to say that there are good guys and bad guys, as the capitalist alienation affects the capitalist too—with the significant difference that he reaps the benefits of the system that alienates him. In any case, we might count the whole of humanity, even the whole of life as the good guys in the losing side. The system is clearly rigged, not only in that it will always work to reproduce and accelerate itself through the violent extraction and accumulation of wealth, but also through the depredation and utter destruction of our natural, cultural and subjective resources.

According to Lacan, capitalism is an awesome thing; a madly accelerating discourse, a cannibal-discourse or even an Ouroboros-discourse:

> The crisis, not of the discourse of the master, but of the capitalist discourse, which is its substitute, is open. I am not saying that capitalist discourse is rotten, on the contrary. It is incredibly clever … but headed for a blowout all the same…. It is the cleverest thing ever devised as a discourse. But it is no less destined to collapse. This is because it is unsustainable … indeed, it

could not go better, but it actually goes too fast. It consummates itself, it consummates itself so well that it consumes itself.[24]

How can we get out of this deadlock?

In that same conference, Lacan talks about a discourse that is still to come, a pestilent form of psychoanalysis, in the sense Freud gave to it when he brought it across the Atlantic for the first time, in 1909.[25] As psychoanalysis follows the footsteps of contemporary art, as it learns from indigenous peoples, feminist movements, decolonial thinking, or queer studies, as it continues to listen to the ignored social and subjective symptoms, as it allows the genuinely pestilent voices to fill our books, our discussions and our practices, it may still have a role to play in this environmental crisis and the subversion of the discourse that begets it ... before it consumes us all.

Georges Didi-Huberman's *Soulèvements* (*Uprisings*) offers one alternative to fight subjective depredation. Contemporary art, as a transformative-performative practice that resides in an active spectator as much as in the artist or their work, is still a space of subversion. Didi-Huberman's work reminds us that desire is still as strong a force as ever, maybe the only force that is still available to bring forth the potency of subjectivity in act, the act of uprising and revolt. By articulating art ranging from Goya's "Caprichos" from 1780, to contemporary pieces, Didi-Huberman builds a narration of insurgency, the violence that irrupts and erupts against power and its sequestration of desire and life. From "elements unleashed," banners that fly against oppression, to "intense gestures"[26] and loudly exclaimed words that make the public square tremble and roar, desire is what allows joyful affections to rally bodies to form a dancing and chanting multitude. Desire is indestructible, its potency recovered, now fully open, in service of difference and singularity. As I said before, Deleuze takes his theory of affections from Spinoza's *Ethics*, where we find potency as absolutely open, absolutely singular. For Spinoza, the possibility to act comes from the distinct affection that characterises each desire, as determined by each object.[27] Following from this, the potency of subjectivity comes from the limit a subject can impose to enjoyment in order to liberate desire.[28] Fisher warns that there is danger in assuming that language alone can give meaning to this potency of desire, a warning that aims to the logocentric heritage of psychoanalysis. Language is central to my understanding of subjectivity and to my practice, but Lacan's RSI and his insistence on topology allow for the inclusion of other registries of experience beyond language.[29]

Psychoanalysis is a *dispositif* of subjectivation through desire. It rises against the constitution of the subject as equivalent to the individual; its weak force resisting normalisation and localisation of all ways of life. It is in this sense that a psychoanalysis can function in favour of the proliferation of life, of its continual fugue—as long as it remains restless, always at the border, being and producing borders. We must ask ourselves if our practice is really a subversive one, if the opening of one individual to the experience of unconscious desire, the practice of psychoanalysis, contributes to a defence of the subject; that is, to the construction of spaces where anxiety, pain, boredom, and sadness are bearable. We must ask ourselves if our practice contributes to the restitution of historisation; if it opens the possibility of a cut that interrupts the seductive vertigo of the markets in order to open other subjectivities, other relations to nature that are not modelled by domination and consumption. This calls for a defence not of the individual, not of the person, but of the subject. To defend it despite the individual, to rescue it from the person. If there is something to be learned from the Freudian adventure it is that giving voice to pain, allowing it to become suffering, can have incredibly powerful consequences.

The practice of psychoanalysis does not happen in a void; it is always a political endeavour. It is a practice that can contribute to the disarticulation of mechanisms of control, particularly in the reproduction of the socially instituted practice of consumption. By doing so, it can open Western and Westernised subjects to the appreciation and care of difference and diversity; it can contribute to weakening the fallacy of self-sufficiency that sustains individualism and the stubborn destruction of "nature" as the material conditions necessary to sustain and reproduce life.

When Leonard Cohen sang the war was over, he also announced that a plague was coming. He might have intended it as a warning of dire things to come. Still, in these desperate times, I give myself licence to believe that this plague, psychoanalysis as a practice of subjective uprisings, a practice that recovers the potency of desire, can contribute not to the preservation of this world, but to its final destruction. As the Zapatistas insist, our world, the world of capitalist depredation, must come to an end if we want a new world: *un mundo donde quepan muchos mundos.*[30]

Notes

1. I think of macro and micropolitics in the way Suely Rolnik presents them in *Esferas de la insurrección: Apuntes para descolonizar el inconsciente*, trans. Cecilia Palmeiro, Marcia Cabrera, and Damian Kraus (Buenos Aires: Tinta Limón, 2019).
2. Juan Camilo Cajigas-Rotundo, "La biocolonialidad del poder: Amazonia, biodiversidad y ecocapitalismo," in *El giro decolonial: Reflexiones para una diversidad epistémica más allá del capitalismo global*, ed. Santiago Castro-Gómez and Ramón Grosfoguel (Bogotá: Siglo del Hombre, 2007), 169–93.
3. Cajigas-Rotundo, "La biocolonialidad del poder," 171.
4. Karl Marx, *Contribución a la crítica de la economía política*, trans. Marat Kuznetsov (Mexico City: Siglo XXI, 2003), 18.
5. For more on a history of nature in Western culture see Cajigas-Rotundo, "La biocolonialidad del poder," 170.
6. Martin Heidegger, "The Question Concerning Technology," in *The Question Concerning Technology and Other Essays*, trans. William Lovitt (New York: Harper & Row, 1977), 3–35.
7. Global Witness, "Enemies of the State?: How Governments and Businesses Silence Land and Environmental Defenders," *Global Witness* (report), July 30, 2019, https://www.globalwitness.org/es/campaigns/environmental-activists/enemies-state/ (accessed May 29, 2020).
8. In 2017 and 2018, I participated in two colloquia, organised by 17, Instituto de Estudios Críticos, Asamblea Nacional de Afectados Ambientales, Unión de Científicos Comprometidos por la Sociedad, and ETC Group. The subject of these gatherings was environmental chaos. What I listened to then became the seeds of this text. See "Coloquios," *17, Instituto de Estudios Críticos*, 2017–2018, https://17edu.org/coloquios/ (accessed September 14, 2019). For more information on these three types of depredation, readers can refer to the works presented there by experts and social leaders.
9. Yásnaya Elena Aguilar, "Los pueblos indígenas no somos la raíz de México, somos su negación constante," *El País*, September 9, 2019,

https://elpais.com/cultura/2019/09/08/actualidad/1567970157_670834.html (accessed May 29, 2020).
10. Cajigas-Rotundo, "La biocolonialidad del poder," 181.
11. Jacques Lacan, "Fonction et champ de la parole et du langage en psychanalyse," in *Écrits* (Paris: Éditions du Seuil, 1966), 245-47.
12. Jacques Lacan, *L'envers de la psychanalyse* (1969–1970), 41 http://staferla.free.fr/S17/S17%20L'ENVERS.pdf (accessed September 14, 2019).
13. Michel Foucault, *Vigilar y castigar: Nacimiento de la prisión*, trans. Aurelio Garzón (Mexico City: Siglo XXI, 1978), 199.
14. Gilles Deleuze, "Post-scriptum sobre las sociedades de control," in *Conversaciones*, trans. José Luis Pardo (Valencia: Pre-textos, 1996), 277. On page 119, Deleuze says "modern power" is what Burroughs calls "control."
15. Manuel Hernández has shown how, after World War II, there was a coordinated effort to produce a new kind of subjectivity that would substitute the citizen in Western democracies: the consumer. See Manuel Hernández. *Subjetivaciones neocoloniales* (Mexico City: Litoral Editores, 2019) ePub.
16. Ignacio Lewkowicz, "Subjetividad adictiva: un tipo psicosocial instituido: Condiciones históricas de posibilidad," in *Las drogas en el siglo… ¿qué viene?* eds. Juan Dobón and Gustavo Hurtado (Buenos Aires: FAC, 1999).
17. For a detailed description of this logic, based on Heidegger's concept of anxiety, see Juan Luis de la Mora, *Pensar la nada: Del consumo a la adicción* (Mexico City: Casa Editorial Abismos, 2016), 63–69.
18. Jacques Lacan, "Le séminaire sur 'La Lettre volée,'" in *Écrits* (Paris: Éditions du Seuil, 1966), 59.
19. See de la Mora, "Una historia de dispositivos y subjetivaciones," *litoral n. 49. El dispositivo psicoanalítico* (Mexico City: Litoral Editores, 2020).
20. Fyodor Dostoyevsky, *Notes from the Underground*, Project Gutenberg, 1996, produced by Judith Boss, http://www.gutenberg.org/files/600/600-h/600-h.htm (accessed September 14, 2019).
21. Gilles Deleuze, *Derrames entre el capitalismo y la esquizofrenia*, trans. Equipo Editorial Cactus (Buenos Aires: Cactus, 2005), 290–92.
22. Andy Fisher, *Radical Ecopsychology: Psychology in the Service of Life* (New York: State University of New York, 2002), 165.

23. When we listen to those who deny global warming, those who oppose a more just society, those in favour of the continued persecution of indigenous peoples and the extractive madness; when we listen to the actual form of discourse, the arguments, the discussions—we are talking about the highest spheres of power, CEOs, bankers, lawmakers, presidents—when we realise the level of blindness, the irrationality, the violence of the *solutions* that are designed and peddled to go to any and all extents to just not stop, it all appears to be a joke. It must be a joke that, facing a global environmental apocalypse, we keep building oil pipes and dams, we keep rescuing banks, and we are willing to develop geo-engineering techniques to counteract the consequences of our actions, but not to stop, never to stop.
24. Jacques Lacan, "Du discours psychanalytique," in *pas-tout Lacan*, http://ecole-lacanienne.net/wp-content/uploads/2016/04/1972-05-12.pdf (accessed September 14, 2019).
25. See Jacques Lacan, "La chose freudienne," in *Écrits* (Paris: Éditions du Seuil, 1966), 403.
26. Descriptions taken from the curatorial texts of Georges Didi-Huberman, "Sublevaciones," Curatorial texts from museum exhibition (MUAC, Mexico City. 2018) https://muac.unam.mx/exposicion/sublevaciones; and http://soulevements.jeudepaume.org/ (accessed September 14, 2019). See also the show's website, http://soulevements.jeudepaume.org/ (accessed September 14, 2019).
27. See Baruch Spinoza, *Ethica, ordine geometrico demonstrata*, 3p56 and scholium, n.d., http://ethicadb.org/pars.php?parid=3&lanid=0#356 (accessed September 14, 2019).
28. Lacan stated that "only love allows enjoyment to condescend to desire." Jacques Lacan, *L'Angoise* (1962–1963), 117, http://staferla.free.fr/S10/S10%20L'ANGOISSE.pdf (accessed September 14, 2019).
29. Fisher, *Radical Ecopsychology*, 63.
30. "A world in which many worlds can be". See EZLN, "Cuarta declaración de la Selva Lacandona," Section III, first published January 01, 1996, http://enlacezapatista.ezln.org.mx/1996/01/01/cuarta-declaracion-de-la-selva-lacandona/. There is an English translation available, although it's quality leaves much to be desired: https://radiozapatista.org/?p=20287&lang=en (accessed July 25, 2020).

Bibliography

Aguilar, Yásnaya Elena. Los pueblos indígenas no somos la raíz de México, somos su negación constante. *El País*. September 9, 2019. Accessed May 29, 2020. https://elpais.com/cultura/2019/09/08/actualidad/1567970157_670834.html.

Anonymous Traditional. 1991. Tlalticpac…. In *Ómnibus de poesía mexicana*, ed. Gabriel Zaid, 52. Mexico City: Siglo XXI.

Cajigas-Rotundo, Juan Camilo. 2007. La biocolonialidad del poder: Amazonia, biodiversidad y ecocapitalismo. In *El giro decolonial: Reflexiones para una diversidad epistémica más allá del capitalismo global*, ed. Santiago Castro-Gómez and Ramón Grosfoguel, 169–193. Bogotá: Siglo del Hombre.

Coloquios. 2017–2018. *17, Instituto de Estudios Críticos*. Accessed September 14, 2019. https://17edu.org/coloquios/.

de la Mora, Juan Luis. 2020. Una historia de dispositivos y subjetivaciones. *litoral* 49. *El dispositivo psicoanalítico.* Mexico City: Litoral Editores.

———. 2016. *Pensar la nada: Del consumo a la adicción.* Mexico City: Casa Editorial Abismos.

Deleuze, Gilles. 2005. *Derrames entre el capitalismo y la esquizofrenia*. Translated by Equipo Editorial Cactus. Buenos Aires: Cactus.

———. 1996. Post-scriptum sobre las sociedades de control. In *Conversaciones*, trans. José Luis Pardo, 227–285. Valencia: Pre-textos.

Didi-Huberman, Georges. 2018. Sublevaciones. In *Curatorial Texts from Museum Exhibition*. Mexico City: MUAC. Accessed September 14, 2019. https://muac.unam.mx/exposicion/sublevaciones; and http://soulevements.jeudepaume.org/.

Dostoyevsky, Fyodor. 1996. *Notes from the Underground.* Project Gutenberg. Produced by Judith Boss. Accessed September 14, 2019. http://www.gutenberg.org/files/600/600-h/600-h.htm.

EZLN. Cuarta declaración de la Selva Lacandona. First published January 01, 1996. Accessed July 25, 2020. http://enlacezapatista.ezln.org.mx/1996/01/01/cuarta-declaracion-de-la-selva-lacandona/. English translation "Fourth Declaration of the Lacandon Jungle," https://radiozapatista.org/?p=20287&lang=en.

Fisher, Andy. 2002. *Radical Ecopsychology: Psychology in the Service of Life.* New York: State University of New York.

Foucault, Michel. 1978. *Vigilar y castigar: Nacimiento de la prisión.* Translated by Aurelio Garzón. Mexico City: Siglo XXI.

Global Witness. Enemies of the State?: How Governments and Businesses Silence Land and Environmental Defenders. *Global Witness* (report). July 30, 2019. Accessed May 29, 2020. https://www.globalwitness.org/es/campaigns/environmental-activists/enemies-state/.

Heidegger, Martin. 1977. The Question Concerning Technology. In *The Question Concerning Technology and Other Essays*, trans. William Lovitt, 3–35. New York: Harper & Row.

Hernández, Manuel. 2019. *Subjetivaciones neocoloniales*. Mexico City: Litoral Editores. ePub.

Lacan, Jacques. Du discours psychanalytique. In *pas-tout Lacan*. Accessed September 14, 2019. http://ecole-lacanienne.net/wp-content/uploads/2016/04/1972-05-12.pdf.

———. 1966a. Fonction et champ de la parole et du langage en psychanalyse. In *Écrits*, 245–47. Paris: Éditions du Seuil.

———. 1966b. La chose freudienne. In *Écrits*, 401–436. Paris: Éditions du Seuil.

———. *L'Angoise*. 1962–1963. Accessed September 14, 2019. http://staferla.free.fr/S10/S10%20L'ANGOISSE.pdf.

———. *L'envers de la psychanalyse*. 1969–1970. Accessed September 14, 2019. http://staferla.free.fr/S17/S17%20L'ENVERS.pdf.

———. 1966c. Le séminaire sur 'La Lettre volée.'. In *Écrits, 11–61. Paris: Éditions du Seuil.*

Lewkowicz, Ignacio. 1999. Subjetividad adictiva: un tipo psicosocial instituido: Condiciones históricas de posibilidad. In *Las drogas en el siglo… ¿qué viene?* ed. Juan Dobón and Gustavo Hurtado. Buenos Aires: FAC.

Marx, Karl. 2003. *Contribución a la crítica de la economía política*. Translated by Marat Kuznetsov. Mexico City: Siglo XXI.

Rolnik, Suely. 2019. *Esferas de la insurrección: Apuntes para descolonizar el inconsciente*. Translated by Cecilia Palmeiro, Marcia Cabrera, and Damian Kraus. Buenos Aires: Tinta Limón.

Spinoza, Baruch. *Ethica, ordine geometrico demonstrata*. Accessed September 14, 2019. http://ethicadb.org/pars.php?parid=3&lanid=0#356.

8

Enjoying the Heat: Anxiety, Fantasy, and Doomsday Prepping

Calum Matheson

In his 2016 book *Facing Climate Change: An Integrated Path to the Future*, climate scientist and Jungian analyst Jeffrey Kiehl urges us to respond in new ways to ongoing climate change. We need to think in grand terms about the future and our individual roles in it, Kiehl seems to say, employing his blend of science, psychoanalysis, and Buddhist mindfulness creating "new stories," even a "new Living Myth," to "turn toward a future of unparalleled potential."[1] In many ways, this approach is not new. First, climate change is presented in all of its potential horror, including but not limited to the general collapse of human society. Next, we are urged to "think big" about solutions and to imagine a new path for ourselves, our communities, our species. Horror alternates with renewal. As Kiehl says, "I sense this deep fear whenever I talk with people about global warming. The seeds of change are embedded in this generalized anxiety."[2]

Human beings have been imagining the end of the world—and their possible salvation—for a long, long time. Apocalyptic myths have existed

C. Matheson (✉)
University of Pittsburgh, Pittsburgh, PA, USA
e-mail: matheson@pitt.edu

C. Burnham, P. Kingsbury (eds.), *Lacan and the Environment*, The Palgrave Lacan Series, https://doi.org/10.1007/978-3-030-67205-8_8

in almost all societies and religions throughout history. The threat of nuclear war after 1945 in American culture presaged the current threat of climate change in many ways. Here too the spectre of apocalypse loomed due to human technological artifice. Here too governments were urged to do more to stave off the worst through international agreements and unilateral actions while individuals turned to spirituality, radical politics, and mass demonstrations. And significantly, a glut of post-apocalyptic fiction imagined the end of the world and its rebirth, alongside scientific warnings about human extinction, as detailed in *Desiring the Bomb: Communication, Psychoanalysis, and the Atomic Age*.[3] As a whole, the imagined solutions were not adequate to the threat: global disarmament never occurred, public fear was easily manipulated, and the nuclear weapons complex still exists worldwide, changed little except in scope and composition. The end of the United States-Soviet nuclear standoff did not come from radical ideas, but rather from the collapse of one competitor. Nuclear fantasies did not trigger anxiety and then change: they mostly triggered more nuclear fantasies. In the context of climate change, harsh warnings about the end of the species have not yet reformed the global economy. We continue to imagine ruined worlds and new futures, alternating with one another. In response to the threat of societal collapse, many of the rich and powerful are already thinking big—salvific imaginations of geoengineering mirror Cold War fantasies like Project Plowshare in which nuclear weapons would have been used *peacefully* to alter Earth's physical geography. Meanwhile, the ultra-ultra-rich quietly develop contingency plans to survive the end and then to rebuild in isolated communes, in abandoned nuclear silos reproducing idealised American communities, in the new frontier of New Zealand, or even on Mars, while less affluent doomsday survivalists become features on television shows like *Doomsday Preppers*. Instead of provoking anxiety, these new imaginations serve as an anchor of enjoyment in the Lacanian sense of *jouissance*. However painful, we are compelled to repeat them in a seemingly endless cycle. Before mindfulness, Carl Jung, and big ideas can save us, we should recognise that fantasies of doom carry a certain appeal, a promised glimpse of the phoenix amidst the ashes, and a chance to peel away the perceived inauthenticity of modern mass culture in favor of new beginnings. Many have used the claim such as "each of us has the capacity to transform our lives,"[4] not to save the neighbours from the rising tide, but to imagine a world in which others are already dead and we can simply start all over.

This chapter explores survivalism as one aspect of response to apocalyptic discourses. My central contention is that it demonstrates the operation of a *fort-da* dynamic, a repeated fantasy of control over presence and absence which serves as an anchor for enjoyment. A danger of lurid descriptions of potential climate catastrophe is that, rather than producing anxieties that lead to collective efforts to preserve Earth, it may enable a fantasy of survival and rebuilding which serves to distance subjects from the urgency of these tasks. The possibility of post-apocalyptic survival is intoxicating because it promises contact with the Real, an impossible access to unmediated reality. Just as much post-apocalyptic American fiction relies on the tropes of Westerns, survivalism encourages us to relate to nature in a heroic, often hyper-masculine mode where we seek less to overcome danger than to revel in the fantasy of a direct encounter with it.[5] To illustrate this argument, I will rely on some dramatic survivalist plans espoused by Silicon Valley executives, precisely those whose resources, money, and social power would be necessary for an effective mitigation of climate change (if indeed such a thing is still possible).

Fort-Da

In *Beyond the Pleasure Principle*, Sigmund Freud recounts his observations of a child (his grandson) whose mother comes and goes while the child is in its crib. The toddler repeatedly throws away and pulls back a toy on a string, making sounds as it does so, that Freud interprets as *fort* (gone) and *da* (here).[6] Although this exercise could be read in a number of ways, a persuasive interpretation within the Lacanian frame is that the child is not so much inuring itself to its mother's absence as it is simulating its control over the conditions of presence and of absence. Faced with a situation where it is powerless—its entire world as represented when the mother's comes and goes without warning—the child is not "toughening up" but performing a compensatory fantasy. The object of enjoyment is not the toy, as the child could keep it or simply throw it away if it so wished. The child instead enjoys its own the sense of its own subjectivity, its capacity to make this small part of its world—which stands in for the whole thing—come and go.[7]

This process of repetition helps to form a subject. The child is first made aware of its separation from the world in a basic sense as the object of its desire can come and go uncontrollably, but it quickly develops a fantasy that assuages this sense of discontinuity. What is often overlooked in this account is the contingent nature of repetition itself. While its iterations form a kind of overall structure, the performance of each individual component (throwing away, bringing back, etc.) seems for the subject a deliberate choice. Thus, the grammar of the subject and its perceived relation to the rest of its Symbolic world and Imaginary fantasy accrete over time: compulsive repetition towards an object girds the fantasy, cements the object's seeming importance in the place of the lack, and defines the subject for itself. This interplay between the repeated acts (such as enunciated speech) and the larger structure of desire might help to explain what Jacques Lacan means when he says that "*Fort-da* is already a rhetorical figure."[8] As individual acts deviate from or conform to the channels of the Symbolic, they help to alter or to reinforce those channels, like water deepening a riverbed or eroding its banks.

In some sense, the *fort-da* process always relies on an aspect of the Real. Although this term is notoriously difficult to define in Lacan's work, at a minimum it describes instances in which the Symbolic fails—instances that are "impossible" within the socially-mediated reality in which we live. The Real is not so much extra-Symbolic existence in the sense of Quentin Meillasoux's *After Finitude* "arche-fossil"[9] as it is a sudden rending of that reality by the intrusion of something that does not fit. Exposing the incompleteness of the Symbolic, however briefly, throws the subject out of joint, as all of its familiar operations seem uncertain. The subject itself is built on a separation from the world. In George Bataille's terms, this is discontinuity—the subject is possible only on the condition that it is something discrete from everything else, and that separation is a product of the entry into language,[10] a perspective similar to Lacan's Mirror Stage, described in *Écrits*.[11] Discontinuous subjects continue to desire an experience of continuity which is impossible because it defies the conditions by which the subject exists in the first place. Separation is also, of course, the condition of desire—we cannot desire something unless it is seen as *other* to some degree. The inevitability of discontinuity establishes the lack in the subject and the subject's desire results from this lack, like an inexhaustible vacuum drawing in proximate objects without ever being filled. The entire signifying network in which a subject operates—language—covers this division, allowing us to act *as if* signifiers could reforge these connections. Of course, Lacan's three orders (Imaginary, Symbolic, and Real) overlap. I want to suggest that the Real helps to animate the Symbolic by investing at least some signifiers with a sense of lost continuity that serves as an anchor for *Besetzung* (in Freud's *Studies on Hysteria*, translated by Strachey as "cathexis").[12] Some signifiers may be more *powerful* than others because they make better promises of the Real. Apocalyptic signifiers are stark examples.

Apocalypse and the Real

As previously mentioned, imaginations of the world's end are not new. In a variety of twentieth-century American cultures, the Bomb served as a condensing symbol for the collapse of human civilisation and its possible extinction while simultaneously serving a function of enjoyment.[13] More recently, all in 1959, nuclear-apocalyptic works like Walter M. Miller Jr's novel, *A Canticle for Liebowitz*, *On the Beach*, by screenplay writer John Paxton, and Pat Frank's novel *Alas, Babylon* have been supplemented with an increasing number of climate-apocalypse tales. The 1995 film, *Waterworld*, written by Peter Rader and David Twohy, was an early entry, widely remembered despite its commercial failure, as were Roland Emmerich and Jeffery Nachmanoff's 2004 film *The Day after Tomorrow* (itself a play on the 1983 anti-nuclear film by Edward Hume, *The Day After*), and more recently, the 2013 film written by Bong Joon-ho and Kelly Masterson, *Snowpiercer*. Ostensibly non-fiction books like Jonathan Schell's 1982 *Fate of the Earth*, have given way to Al Gore's 1992 book, *Earth in the Balance*, his 2006 documentary, *An Inconvenient Truth*, and their like. Media outlets frequently report climate change in apocalyptic terms as well, and popular movements like Extinction Rebellion rely on this message for organisation.

These fantasies of global destruction do not bring us face-to-face with the Real of extinction. Even in the most extreme visions—like Alan Weisman's 2007 book and series *The World Without Us*—some perspective is required to depict the aftermath. This claim echoes one by Freud, who argued that we cannot imagine our own deaths. The phrase "world without us" echoes a joke mentioned by Alenka Zupančič in *What is Sex?* in which a café patron asks for coffee without cream. The server responds that the café does not have cream but does have milk, so would the customer like coffee without milk?[14] The point of the joke is that the absence still defines the thing—as it does with efforts to imagine extinction. Why so many depictions of ruined human buildings? Why focus on the places well known to us now, and not, say, the chemical environments of microbiota? Even were we to succeed in depicting a post-human world from another perspective, the imagination happens now, in the Symbolic stew of living human culture, and insists that we imagine an opposite case in which we survive.

Apocalyptic imaginations are in some ways attractive. Who does not sometimes wonder, as Frederick Jessett wrote many years ago, "if our interest is so much in survival as in the fascinating possibility we might be legitimately able to get rid of our neighbors."[15] Contemporary survivalism is increasing in both scope and visibility, as the examples in this chapter illustrate. Although the desire not to die in a future environmental crisis may explain part of this, it does not tell the whole story. Many survivalists imagine living off the land when civilisation collapses, by hunting, fishing, and foraging. They imagine far less complicated existences—this is what it means for civilisation to collapse, after all. No taxes to pay. No PTA meetings to attend. No TPS reports to file. Especially in an American context, these fantasies reactivate many of the tropes in classic Westerns. Rugged, hyper-masculine men (and less frequently, women) take the law into their own hands, living in stark (hot, dusty, desertified …) environments exposed to Nature's wrath, living authentic, if violent, lives without the technological sophistication of urban twenty-first-century environments. For survivalists who stockpile ammunition, grow their own food, practice escape routes, and spend austere periods in the wilderness, at some level this is not about overcoming resistance. It is about *encountering* resistance. The imagination of a devastated world in which the subject still lives—but without a hyper-mediated society—is one aspect of a longing for the Real. These fantasies are anchors for enjoyment precisely because when the weekend in the woods is over, one can pack up and go back home. The toy gets pulled back on its string, and *fort* becomes *da* again.

Presence also exists *within* fantasies of destruction outside their contrast to the current world of the living. In post-apocalyptic imaginations that do not include human extinction, societies are generally rebuilt. The destruction of the old order creates the possibility for a new one. Often, these are violent dreams in which a group of "good-guy" homesteaders stand up to "evil" roving gangs, again recreating tropes of civilisation (cultivation, democracy, conservative social norms) against savagery (nomadism, cannibalism, idolatry) familiar from Westerns. Here too, the environment figures powerfully in apocalyptic fantasy. Climatc wastelands are often thought of as "deserts," a landscape that signifies extremity, hardship, and spiritual renewal.[16] The imagined emptiness and hostility of post-climate change environments is not accidental in these fantasies: rebuilding lacks an epic quality without deprivation and hardship, just as the myth of tough, independent American frontier settlers requires resistance to establish heroism. The inhospitability of the environment is an anchor for enjoyment, therefore, partly because it allows climate apocalypticism to serve as a repetition of other destruction/regeneration fantasies that fuse destruction and the rebirth of somehow more pure civilisation. The specific content of rebuilding in these fantasies is less important than the fact that they exist at all—the possibility of destruction is always twinned with rebirth, as it has been since the Revelation of John.

Silicon Valley in the Shadow of Death

Doomsday prepping has traditionally been associated with rural, conservative, white men. Whether or not this demographic assumption was at some point accurate, at least the politics and locations of prepping have begun to change. Silicon Valley, less a place and more of a culture according to one of its boosters,[17] has caught the bug, as a number of recent mass media articles have claimed. Prepping amongst tech company personnel—especially the CEOs—challenges some older conceptions. Although the politics of Silicon Valley are sometimes hard to pin down, prepping seems to have adopted a more liberal veneer.[18] Amongst the ultra-rich, New Zealand features prominently. Although hardly a liberal, billionaire venture capitalist Peter Thiel has apparently developed extensive plans for doomsday, including an escape to the island country, where he has bought land.[19]

Silicon Valley prepping is not entirely motivated by climate change. It differs from traditional survivalism in that what many CEOs fear is something that they themselves have helped to bring about. Automation (assisted by advanced technology) results in increasing unemployment, and that undermines stability; meanwhile, production (again, assisted by, even of, technology) accelerates climate change. Instead of fearing the prospect of avoidable interstate warfare, many fear those who will ultimately be disenfranchised by what Silicon Valley has put in motion.[20] This cognitive dissonance sometimes coexists with the *kill thy neighbours* philosophy of Cold War shelter ethics. One venture capitalist, for example, reportedly keeps guns in case of doomsday "zombies."[21] Such a term may describe the deprived masses of the future in shallow terms reminiscent of Daniel Schreber's "hastily improvised men," but it also has an anti-black racial implication.[22]

The most dramatic difference between Silicon Valley preparations and traditional doomsday prepping has to do with the scale of fantasies for escape and survival. Besides Peter Thiel, who has been described as "Less a person than a shell company for a diversified portfolio of anxieties about the future,"[23] other executives have imagined radical futures in which humanity has overcome the spectre of environmental collapse. Jeff Bezos, CEO of Amazon, has cited climate change as one reason that humanity has to shift industry into outer space[24] because human beings are "destroying this planet."[25] In some ways, everything Elon Musk does seems to be aimed at averting human extinction through radical technological change, from electric cars to space colonisation. Musk imagines both the danger of fossil fuels and the possibility of global war as an impetus for change, but also sees the possibilities: "Mars will need everything from iron foundries to pizza joints."[26] Technological and entrepreneurial optimism exist side-by-side with imaginations of a ruined Earth bereft of human life.

Utopian thinking existed in the context of nuclear weapons, too, perhaps nowhere more clearly than in Project Plowshare, described by Scott Kirsch's *Proving Grounds* as a scheme to use nuclear weapons for a variety of engineering feats like forming new ports in Alaska or blasting an alternative to the Panama Canal.[27] In the context of climate change, one instead finds geoengineering and dramatic climate "moonshots." Geoengineering proposals, increasingly gaining popularity as a whole, have ranged from pouring billions of white balls into the oceans or building giant mirrors in space to deliberately release sulphur particulate pollution into the upper atmosphere.[28] Amongst Silicon Valley elites, reactions to the threat of climate change have tended to veer towards grand plans and away from smaller, practical measures.[29] Scientific feasibility aside, these grand proposals display a salvific impulse that matches dystopian tales in magnitude, simply reversing the sign.

Not-Without-Survival

Apocalyptic climate imaginations, plans for doomsday, and salvific fantasies of technological redemption fit the cycle of *fort-da*. The world is imagined as gone with the survivors testing their mettle against the Real, and then rebuilding a new society. Or, against this alternative, the aristocracy of Silicon Valley paints utopian futures of space colonisation and a geo-engineered Earth. The enjoyment of the sense of control explains the repetition of these various fantasies. In the geoengineering and survival fantasies, it is up to the subject to make new worlds. In the climate-dystopia image, it is still human beings who have destroyed the Earth. It is in our hands to make and unmake the world. At a more quotidian level, whether to indulge in these fantasies at all permits a certain element of enjoyment. We are free to ignore all the warning signs of climate change and even take perverse pride in our ignorance, as some Americans choose to do. Psychoanalytic accounts of culture cannot predict the attitudes of every given individual, of course, but they do suggest that the narrative of apocalyptic imaginations resulting in anxiety, and anxiety spurring change, may be wrong—or even worse, backwards.

The fear sparked by depictions of climate change cannot possibly be an effective spur for action for most subjects. Millions of people saw *The Day After Tomorrow*, for example, and the majority did not become radical environmentalists as a result (the spectacle of Americans fleeing across the southern border in the film should have special resonance now in a time where "build that wall" has become a populist rallying cry). The idea that pop-culture climate fantasies are enjoyable is hardly a leap: after all, people pay to see them, and then when the film is over, they simply leave the theatre or stop watching and do something else. Fear can be exciting, but *enjoyment* results from our simulated control over it. This kind of *fort-da* process does not produce anxiety; but serves as the primary mechanism to keep it away. In his seminar *Anxiety*, Lacan formulates this affect as "*not without an object*," rather than simply objectless.[30] Anxiety does not result from "exposure" to an object as long as we retain a sense of control. It results only when the fantasy of *fort-da* breaks down and the object—in this case, a future devastated by climate change—becomes seemingly unavoidable. Anxiety is a signal of the Real which by definition cannot be mediated in the Symbolic order of language or affixed to as a fantasy.[31]

This dynamic has important implications for how we think about our responses to climate change. Survivalism and technological optimism may seem like opposite ends of a spectrum, but in fact both of them serve as elements in a broader cultural fantasy of control. Survivalists from Silicon Valley to those in the poorest rural communities may in some sense ultimately be *optimists*. Not only do they imagine their own endurance through the next apocalypse, they may also plan to build new, better worlds in line with their fantasies. They may imagine themselves something like Noah, rebuilding the world after its sins have been washed away by the Flood, ready to start again without the errors of the old world. Those who believe that technology will save Earth from its doom imagine that growth can continue forever, all problems to be solved by their very own sources. The seeming inconsistency of the Silicon Valley prepper mindset is thus not a contradiction at all. It represents radically different fantasies that may or may not be expressed by the same tech magnates; but are linked by their distancing effect from the anxiety-producing Real of potential human extinction.

Conclusion: A Place for Anxiety

The dynamics of *fort-da* may help to explain (along with many other economic, political, and psychological factors) why information on the threat of climate change has not yet motivated effective action. This is not to say that anxiety, as Kiehl argues, has no place. A confrontation with the fantasy object that seems too proximate intimates a feeling of encounter with the Real. Sometimes the result is "acting out" in non-productive ways, but there may also be potential in what Ira Chernus's *Dr. Strangegod* calls "going to pieces," the traumatic encounter with despair in relation to possible extinction.[32] Instead of pushing down despair through constant attempts to bolster our subjectivity in the face of an encroaching climate change Real, we should allow the spectre of extinction to dislocate us, pulling apart the fantasies that moor us in social formations where *fort-da* narratives soothe us into potential death. In Lacanian terms, perhaps this entails "subjective destitution" and the attendant move to "ridiculise" not "our knowledge" but the fantasy of control.[33] Or, following Slavoj Žižek's ideas in *Violence*,[34] perhaps we should push back against the chain of repetition that forms when we are always compelled to act immediately in the spectre of some tragedy and instead take the time for critical reflection on the causes of violence.

The ultimate problem that is elided by the *fort-da* of apocalypse, prepping, and utopian optimism is the objective groundlessness of human existence itself.[35] The horror that results from our knowledge that we can—and may—destroy ourselves is an instigator of anxiety in a true sense as a confrontation with the Real, rather than as a diminished feeling of worry that contributes easily to sites of subjective enjoyment. Climate change may or may not be an existential threat. We may or may not be able to arrest its progress. We may or may not be wiped out by an asteroid on the day that our new breakthrough technologies make fossil fuels obsolete. Does it matter? *That* anxiety, resulting from the uncertainty that there is any Other at all and that it may not need or even want us, may be the anxious despair that is a condition of possibility for a future … if we desire one.

Notes

1. Jeffrey T. Kiehl, *Facing Climate Change: An Integrated Path to the Future* (New York: Columbia University Press, 2016), 142
2. Kiehl, *Facing Climate Change*, xi
3. Calum L. Matheson, *Desiring the Bomb: Communication, Psychoanalysis, and the Atomic Age* (Tuscaloosa: University of Alabama Press, 2019), 102–6.
4. Kiehl, *Facing Climate Change*, 142.
5. For a discussion of a similar structure of fantasy in relation to the Real and its mediation in another context, see Jesse Proudfoot and Paul Kingsbury, "Periscope Down! Charting Masculine Sexuation in Submarine Films," in *Psychoanalytic Geographies*, ed. Paul Kingsbury and Steve Pile (New York: Routledge, 2016), 241–43.
6. Sigmund Freud, *Beyond the Pleasure Principle*, trans. James Strachey (New York: W. W. Norton, 1961), 13–17
7. Matheson, *Desiring the Bomb*, 16
8. Jacques Lacan, "Columbia University: Lecture on the Symptom," in *Applied Lacanian Psychoanalysis: Culture/Clinic 1*, ed. Jacques-Alain Miller and Maire Jaanus (Minneapolis: University of Minnesota Press, 2013), 13; emphasis in the original.
9. Quentin Meillassoux, *After Finitude: An Essay on the Necessity of Contingency*, trans. Ray Brassier (London: Continuum Publishing, 2009), 14.
10. Georges Bataille, "Hegel, Death and Sacrifice," *Yale French Studies* 78 (1990): 9–28, https://www.jstor.org/stable/2930112.
11. Jacques Lacan, *Écrits: The Complete Edition in English*, trans. Bruce Fink (New York: W. W. Norton, 2006), 75–81.
12. See Sigmund Freud, *Studies on Hysteria*, trans. James Strachey (London: Hogarth Press and the Institute of Psycho-Analysis, 1955), 89.
13. Matheson, *Desiring the Bomb*.
14. Alenka Zupančič, *What is Sex?* (Cambridge, MA: MIT Press, 2017), 48.
15. Frederick Jessett, "Shelters and Self." Letter to *Christian Century*, December 13, 1961, 1496.
16. See Belden C. Lane, *The Solace of Fierce Landscapes: Exploring Desert and Mountain Spirituality* (New York and Oxford: Oxford University Press, 1998); and Patricia Nelson Limerick, *Desert Passages: Encounters with the American Deserts* (Albuquerque: University of New Mexico Press, 1985), on deserts landscapes; and on colonial nuclearism see Valerie L. Kuletz,

Tainted Desert: Environmental and Social Ruin in the American West (New York: Routledge, 1998).
17. Vinod Khosla, "The Silicon Valley Culture," *Medium*, January 16, 2018, https://medium.com/@vkhosla/the-silicon-valley-culture-bdc86db0b524 (May 30, 2020).
18. Rachel Riederer, "Doomsday Goes Mainstream," *Dissent* (Spring 2018), https://www.dissentmagazine.org/article/doomsday-goes-mainstream-liberal-preppers-silicon-valley (accessed May 30, 2020).
19. Mark O'Connell, "Why Silicon Valley Billionaires are Prepping for the Apocalypse in New Zealand," *Guardian*, February 15, 2018, https://www.theguardian.com/news/2018/feb/15/why-silicon-valley-billionaires-are-prepping-for-the-apocalypse-in-new-zealand (accessed May 30, 2020).
20. Evan Osnos, "Doomsday Prep for the Super-Rich," *New Yorker*, January 30, 2017, https://www.newyorker.com/magazine/2017/01/30/doomsday-prep-for-the-super-rich (accessed May 30, 2020); Jon Henschen, "Tech Exec Doomsday Preppers: Protecting Themselves from the Automation Monster They Created," *Intellectual Takeout*, February 22, 2018, https://www.intellectualtakeout.org/article/tech-exec-doomsday-preppers-protecting-themselves-automation-monster-they-created (May 30, 2020).
21. Olivia Carville, "The Super Rich of Silicon Valley Have a Doomsday Escape Plan," *Bloomberg*, September 5, 2018, https://www.bloomberg.com/features/2018-rich-new-zealand-doomsday-preppers/ (accessed May 30, 2020).
22. Eric King Watts, "Postracial Fantasies, Blackness, and Zombies," *Communication and Critical/Cultural Studies* 14, no. 4 (2017): 317–33, https://doi.org/10.1080/14791420.2017.1338742.
23. O'Connell, "Why Silicon Valley Billionaires are Prepping."
24. Catherine Clifford, "Jeff Bezos: I spend my billions on space because we're destroying Earth," *CNBC*, July 17, 2019, https://www.cnbc.com/2019/07/17/why-jeff-bezos-spends-billions-on-space-technology.html (accessed May 30, 2020).
25. Isobel Asher Hamilton, "Jeff Bezos says space travel is essential because we are 'in the process of destroying this planet,'" *Business Insider*, July 18, 2019, https://www.businessinsider.com/jeff-bezos-space-travel-essential-because-destroying-planet-2019-7 (May 30, 2020); for parallels between the rhetoric of space exploration and the frontier myth see Linda Billings, "Overview: Ideology, Advocacy, and Spaceflight: Evolution of a Cultural

Narrative," in *Societal Impact of Spaceflight*, ed. Steven J. Dick and Roger D. Launius (Washington, DC: National Aeronautics and Space Administration Office of External Relations History Division, 2007), 483–84.
26. Jonathan Shieber, "Elon Musk Is the Ultimate Doomsday Prepper," *Tech Crunch*, March 12, 2018, https://techcrunch.com/2018/03/12/elon-musk-is-the-ultimate-doomsday-prepper/ (accessed May 30, 2020).
27. Scott Kirsch, *Proving Grounds: Project Plowshare and the Unrealized Dream of Nuclear Earthmoving* (New Brunswick, NJ: Rutgers University Press, 2005).
28. Fred Pearce, "Geoengineer the Planet? More Scientists Now Say It Must Be an Option," *Yale Environment* 360, May 29, 2019, https://e360.yale.edu/features/geoengineer-the-planet-more-scientists-now-say-it-must-be-an-option (accessed May 30, 2020).
29. Janko Roettgers, "Big Tech Has Some Wild Ideas to Save the Planet, but Will They Really Work?" *Variety*, September 10, 2019, https://variety.com/2019/digital/news/jeff-bezos-tesla-robert-downey-jr-silicon-valley-climate-change-1203329012/ (accessed May 30, 2020).
30. Jacques Lacan, *The Seminar of Jacques Lacan, Book X: Anxiety, 1962–1963*, ed. Jacques-Alain Miller, trans. A. R. Price (Cambridge: Polity Press, 2014), 89; emphasis in the original.
31. Lacan, *Book X: Anxiety*, 160.
32. Ira Chernus, *Dr. Strangegod: On the Symbolic Meaning of Nuclear Weapons* (Columbia: University of South Carolina Press, 1986), 158–59.
33. Jacques Lacan, "Proposition of 9 October 1967 on the Psychoanalyst of the School," trans. Russell Grigg, 6–7, http://iclo-nls.org/wp-content/uploads/Pdf/Propositionof9October1967.pdf (accessed May 30, 2020).
34. Slavoj Žižek, *Violence: Six Sideways Reflections* (New York: Picador, 2008), 6–8.
35. Matheson, *Desiring the Bomb*.

Bibliography

Bataille, Georges. 1990. Hegel, Death and Sacrifice. *Yale French Studies* 78: 9–28. https://www.jstor.org/stable/2930112.

Billings, Linda. 2007. Overview: Ideology, Advocacy, and Spaceflight: Evolution of a Cultural Narrative. In *Societal Impact of Spaceflight*, ed. Steven J. Dick and Roger D. Launius, 483–500. Washington DC: National Aeronautics and Space Administration Office of External Relations History Division.

Carville, Olivia. 2018. The Super Rich of Silicon Valley Have a Doomsday Escape Plan. *Bloomberg*. September 5. Accessed May 30, 2020. https://www.bloomberg.com/features/2018-rich-new-zealand-doomsday-preppers/.

Chernus, Ira. 1986. *Dr. Strangegod: On the Symbolic Meaning of Nuclear Weapons*. Columbia: University of South Carolina Press.

Clifford, Catherine. 2019. Jeff Bezos: I Spend My Billions on Space Because We're Destroying Earth. *CNBC*. July 17. Accessed May 30, 2020. https://www.cnbc.com/2019/07/17/why-jeff-bezos-spends-billions-on-space-technology.html.

Freud, Sigmund. 1961. *Beyond the Pleasure Principle*. Translated by James Strachey. New York: W. W. Norton.

———. *Studies on Hysteria*. Translated by James Strachey. London: Hogarth Press and the Institute of Psycho-Analysis, 1955.

Jessett, Frederick. Shelters and Self. Letter to *Christian Century*. December 13, 1961, 1496.

Hamilton, Isobel Asher. 2019. Jeff Bezos Says Space Travel Is Essential Because We Are 'in the Process of Destroying This Planet.' *Business Insider*. July 18. Accessed May 30, 2020. https://www.businessinsider.com/jeff-bezos-space-travel-essential-because-destroying-planet-2019-7.

Henschen, Jon. 2018. Tech Exec Doomsday Preppers: Protecting Themselves from the Automation Monster They Created. *Intellectual Takeout*. February 22. Accessed May 30, 2020. https://www.intellectualtakeout.org/article/tech-exec-doomsday-preppers-protecting-themselves-automation-monster-they-created.

Khosla, Vinod. 2018. The Silicon Valley Culture. *Medium*. January 16. Accessed May 30, 2020. https://medium.com/@vkhosla/the-silicon-valley-culture-bdc86db0b524.

Kiehl, Jeffrey T. 2016. *Facing Climate Change: An Integrated Path to the Future*. New York: Columbia University Press.

Kirsch, Scott. 2005. *Proving Grounds: Project Plowshare and the Unrealized Dream of Nuclear Earthmoving*. New Brunswick, NJ: Rutgers University Press.

Kuletz, Valerie L. 1998. *Tainted Desert: Environmental and Social Ruin in the American West*. New York: Routledge.

Lacan, Jacques. 2013. Columbia University: Lecture on the Symptom. In *Applied Lacanian Psychoanalysis: Culture/Clinic 1*, ed. Jacques-Alain Miller and Maire Jaanus, 8–16. Minneapolis: University of Minnesota Press.

———. 2006. *Écrits: The Complete Edition in English*. Translated by Bruce Fink. New York: W. W. Norton.

———. Proposition of 9 October 1967 on the Psychoanalyst of the School. Translated by Russell Grigg, 1–11. Accessed May 30, 2020. http://iclo-nls.org/wp-content/uploads/Pdf/Propositionof9October1967.pdf.

———. 2014. *The Seminar of Jacques Lacan, Book X: Anxiety, 1962–1963*. Edited by Jacques-Alain Miller. Translated by A.R. Price. Cambridge: Polity Press.

Lane, Belden C. 1998. *The Solace of Fierce Landscapes: Exploring Desert and Mountain Spirituality*. New York and Oxford: Oxford University Press.

Limerick, Patricia Nelson. 1985. *Desert Passages: Encounters with the American Deserts*. Albuquerque: University of New Mexico Press.

Matheson, Calum L. 2019. *Desiring the Bomb: Communication, Psychoanalysis, and the Atomic Age*. Tuscaloosa: University of Alabama Press.

Meillassoux, Quentin. 2009. *After Finitude: An Essay on the Necessity of Contingency*. Translated by Ray Brassier. London: Continuum Publishing.

O'Connell, Mark. 2018. Why Silicon Valley Billionaires Are Prepping for the Apocalypse in New Zealand. *Guardian*. February 15. accessed May 30, 2020. https://www.theguardian.com/news/2018/feb/15/why-silicon-valley-billionaires-are-prepping-for-the-apocalypse-in-new-zealand.

Osnos, Evan. 2017. Doomsday Prep for the Super-Rich. *New Yorker*, January 30. Accessed May 30, 2020. https://www.newyorker.com/magazine/2017/01/30/doomsday-prep-for-the-super-rich.

Pearce, Fred. 2019. Geoengineer the Planet? More Scientists Now Say It Must Be an Option. *Yale Environment* 360. May 29. Accessed May 30, 2020. https://e360.yale.edu/features/geoengineer-the-planet-more-scientists-now-say-it-must-be-an-option.

Proudfoot, Jesse, and Paul Kingsbury. 2016. Periscope Down! Charting Masculine Sexuation in Submarine Films. In *Psychoanalytic Geographies*, ed. Paul Kingsbury and Steve Pile, 241–256. New York: Routledge.

Riederer, Rachel. 2018. Doomsday Goes Mainstream. *Dissent* (Spring). Accessed May 30, 2020. https://www.dissentmagazine.org/article/doomsday-goes-mainstream-liberal-preppers-silicon-valley.

Roettgers, Janko. 2019. Big Tech Has Some Wild Ideas to Save the Planet, But Will They Really Work? *Variety*. September 10. Accessed May 30, 2020. https://variety.com/2019/digital/news/jeff-bezos-tesla-robert-downey-jr-silicon-valley-climate-change-1203329012/.

Shieber, Jonathan. 2018. Elon Musk Is the Ultimate Doomsday Prepper. *Tech Crunch*. March 12. Accessed May 30, 2020. https://techcrunch.com/2018/03/12/elon-musk-is-the-ultimate-doomsday-prepper/.

Watts, Eric King. 2017. Postracial Fantasies, Blackness, and Zombies. *Communication and Critical/Cultural Studies* 14 (4): 317–333. https://doi.org/10.1080/14791420.2017.1338742.

Žižek, Slavoj. 2008. *Violence: Six Sideways Reflections*. New York: Picador.

Zupančič, Alenka. 2017. *What Is Sex?* Cambridge, MA: MIT Press.

Part III

Lack of Knowledge

9

Does the Animal Desire?

Alessandra Capperdoni

Speech is essentially the means of gaining recognition.... That is why, in a sense, one can speak of the language of animals. Animals have a language [langage] to the degree that there is someone there to understand it.
—Lacan, *Seminar Book I, Freud's Papers on Technique*

Does the animal desire? And what does the animal want? The question, which has lingered in enduring yet submerged ways in the history of human thought, rises spontaneously anew at a time when the relationship between the human and the other-than-human natural world is examined from many fronts—political and environmental activists, animal rights defenders, philosophers, cultural workers, and, quite simply, animal lovers. It is also a question that opens a Pandora box, for the use of terms such as *animal* and *nature* bring along a thousand-year old tradition in Western and Non-Western thought that is, to say the least, overdetermined. This paper does not have the ambition to solve the

A. Capperdoni (✉)
Simon Fraser University, Burnaby, BC, Canada
e-mail: acapperd@sfu.ca

C. Burnham, P. Kingsbury (eds.), *Lacan and the Environment*, The Palgrave Lacan Series, https://doi.org/10.1007/978-3-030-67205-8_9

question. Instead, it is interested in refocusing the "question of the animal" (the anthropocentric discourse about animality) through the lens of Lacanian psychoanalysis, and asks whether we can speak of desire "in the animal," how such desire is constituted in relation to the human, and what ethical questions it poses for the human subject. In so doing, it will ask whether such (re)placement of the human-animal relation can be read through Jacques Lacan's notion of the discourse of the analyst. If we can speak of an "environmental turn" in the Humanities, these questions will hopefully address a lack in Lacanian thinking (no pun intended), that is, a thinking about the animal that does not reduce animality to the gaze of the human subject but rather carries forth what Lacan himself, despite his reservations, seems to have gestured at in the last years of his teaching: the fundamental kinship, in their irreducible singularities and non-relationality, of the "human" and "animal" worlds.[1] In my discussion, I am indebted to the many thinkers who time and again have come back to the question of the animal in relation to psychoanalysis and, to quote Peter Buse, to what the animal is *for* Lacan.[2] This is a slippery terrain of inquiry and one can only be grateful to those who attempted to tread it, albeit tentatively. But despite Lacan's clear critical detachment of psychoanalysis from the tradition of philosophy in his later work (a detachment which cannot erase the longstanding engagement of Lacan with many philosophers of the continental tradition), this discussion will also need to address the time-old question of the animal in philosophical discourse (that reservoir of the symbolic register) in order to consider, in particular, what phenomenology and deconstruction can offer to Lacanian discourse—or, perhaps, how the "margins of philosophy" can speak to Lacanian psychoanalysis.[3] It is with the latter that I am going to begin my discussion.

Is the Animal that Enjoys the Animal that Speaks?

In the talks given at Sainte-Anne in 1971-1972, later transcribed in the unedited collection *The Knowledge of the Psychoanalyst*, Lacan goes back to some of the tenets of his teaching in order to discuss the knowledge of the analyst.[4] Two of these are particularly appropriate to address the conundrum at the heart of the animal question through the discourse of psychoanalysis: the position of speech and the body in relation to *being* and enjoyment. The first tenet highlights the centrality of the function of speech in the human subject and, subsequently, the significance of speech in the process of analysis: "It is from being speaking—I apologize for the first being—that he comes to being, *anyway that he has this feeling. Naturally he does not get to it—he fails.* But this dimension of being that has opened up all of a sudden, one could say that for a good period of time, it had an effect on the system … of philosophers at least."[5] For Lacan the subject is a speaking subject and possibilities of analytic intervention are always at the level of speech. The second tenet regards the position of the body: "Where does enjoyment lie? What does it need? A body. To enjoy, a body is necessary."[6] I will come back to the implications of these notions and how Lacan embeds them in a larger discussion on knowledge and truth. For the time being, they seem to be good entry points in thinking about the way in which the question of the animal has functioned in philosophical discourse but also the way in which philosophy is always implicated with the discourse of psychoanalysis.

Speech [*parole*], which is impingent on the possibilities of language [*langue*], and the body constitute fraught notions in the history of Western thought. Notably, Aristotle defines the human as *zoon logon echon*, the animal endowed with language, hence the rational animal.[7] This means that the mediation of language is a necessary condition to access the humanity of the human. To its animality, there needs something to be added in order to have Man. Aristotle's formulation provides a paradigm through which the future centuries of Western thought were to conceptualise the human in relation to language. From the scale of being of medieval thought—ranging from the lower level of elemental entities and the living body of animals (the sentient) to the essence of Man (the rational) created in the image of the Divine through his Word—to René Descartes's foundation of modern rationalism in the seventeenth century, the animal is not only relegated ontologically to the lower echelon of the living but constituted as the necessary negative condition of "being Man": the animal being *is* "lack" because it lacks language and reason. Descartes's formulation of radical doubt about experience and the sensory perception outside of thought (*Ego cogito, ergo sum*) places itself in this tradition despite the ruptures it effects.[8] Not surprisingly, Descartes's mechanistic theory labels animals as *automata* in view of their assumed lack of reason and speech. Animals react to stimuli as mere machines that cannot think through, and thus cannot truly perceive and express as thought, their pain—a notion that de facto justifies vivisection. One century later, Immanuel Kant makes of the animal a source of ethics for the human: cruelty towards the animal is not unjustifiable *per se* but because it enables the possibility of human infliction of pain onto other humans. For Kant, the animal does not have access to subjectivity, for it is only through self-consciousness—the awareness of the mind's ability to construct a unified experience—that the subject comes into being. Animals, for Kant, may have desires but, unlike humans, they lack the "will"—that is, they lack the ability to stand back and choose the subsequent action to such desires. The consequences of such thought are not much different from Descartes': "Our duties towards animal, then, are indirect duties toward mankind."[9] This implicit anthropocentrism was not to be abandoned in later critiques of Cartesian and Kantian thought. In the *Philosophy of Nature*, Georg Wilhelm Friedrich Hegel's animality is implicated in his discussion of disease, and in the *Realphilosophie* (or Jena Lectures) that precedes the *Phenomenology of Spirit* by a year, Man is defined as the sick animal, for this is the price to pay for his humanity: with the emergence into existence natural life is abandoned.[10] For Hegel, the animal is the inert object at home in the world. The effort to reappropriate his full self-consciousness can only be Man's work![11] The notion of man as the sick animal was later revived by Friedrich Nietzsche himself. But Nietzsche's critique of Hegelian thought and sympathy for animal life is also not untinged with anthropocentrism.[12]

Amongst these voices we also find notable exceptions. The Greek Pythagoras and Plutarch were avowed vegetarians who were obsessed with human cruelty towards animals and stood for the idea of the kinship of the living. Michel de Montaigne takes issue with Cartesian thought and asks, in his "An Apology of Raymond Sebond," "How does he [Man] know, by the force of his intelligence, the secret internal stirrings of animals?… When I play with my cat, who knows if I am not more a pastime to her than she is to me?"[13] And while Voltaire critiques the contradictory discourses of modern philosophers about animality and David Hume claims the "'evident truth' that animals are endowed with 'thought and reason,'"[14] Jeremy Bentham shifts the discourse altogether: "the question is not, Can they *reason*? nor, Can they *talk*? but, Can they *suffer*?"[15]

One has to ask whether Montaigne and Bentham, in stepping outside of the normative discourse of philosophy, are occupying the place of the analyst interrogating the (divided) subject of philosophy—divided because unable to account for what is in excess of what is constructed as reason or self-consciousness. If this is the case, their discourse is opening up to the production of a new master signifier and a "differend" truth.[16] What their claims bring forth with respect to the animal-question is the animal-subject's relationship to the Other and the place of the body in enjoyment/pain. In fact, neither Montaigne nor Bentham speak about "the animal" in the singular but only about animal*s*. For such animals, pleasure and pain are not the exclusive domain of biology (they are reduced neither to instinct nor mechanistic reaction to stimuli) but are constituted through the relation with other animals—including the human-animal—and the symbolic that frames these relations.

It is perhaps premature to speak about the analyst's discourse as discussed by Lacan in *Seminar XVII* in the context of these counter-voices of modern Western philosophy, though we can talk of psychoanalytic intervention.[17] Neither of these thinkers was to develop a full discourse about animality, but it is important to notice that the dissidence of their positions interrupts the progressive (albeit anxious) construction of the human in modern thought as the non-animal or the surplus of animality—in which Giorgio Agamben reads a central condition of the "anthropological machine" of Western thought.[18] Montaigne and Bentham's speculations constitute important interventions and dislodgment of such thought. Montaigne's cat reverses the epistemology of subject and object, showing that only from an anthropocentric perspective humanity is extensive with subjectivity. To the cat, the human is also Other (a relation to otherness that Emmanuel Levinas will reframe by introducing the function of the face in his *Ethics and Infinity*), not only because it is an "other" being altogether, but because the cat has to contend with the order of the symbolic that the human imposes upon her (the register of the symbolic also implicates animals' lives in human lives).[19] In similar fashion, Bentham unsettles Cartesian certitudes about the animal's mechanistic reflex by drawing attention to a suffering that may be biologically determined by the presence of a complex nerve system but does not have to imply a lack of awareness. In his *Descent of Man* (1871), Charles Darwin was to cite a pertinent example: the dog who during vivisection pleads with its torturer by licking the hand that holds the knife speaks of a scene of address of the animal to the human.[20] Although the *quality* or intensity of such suffering is difficult to demonstrate, we cannot strip down the pain of the dog to mechanistic reflex, for its pleading involves the Other (now master and perpetrator of violence) in demanding, quite simply, *please, stop killing me!*

Does the licking of the hand constitute a scene of address? Does it amount to *speech* and *demand*? As Eduardo Kohn notes, "nonhuman life forms also represent the world" and represent "us." If this is the case, seeing, knowing, and thinking are "not exclusively human affairs."[21] In a short article appeared in *Difficiles libertés* in 1963, Levinas recalls the memory of Bobby, the dog who every day eagerly welcomed the arrival of the prisoners at the end of their labour in the Nazi camp of Fallingsbostel.[22] In musing on the discourses about humanity and animality, Levinas notes how in this space of death and non-recognition—where the prisoners are stripped of their own humanity by gaze of the Other, deprived of any acknowledgement of their humanity ("We were subhuman, a gang of apes"), beings without language, "entrapped in their species"—it is in the encounter with the animal that they regain their human subjectivity. The dog that the prisoners call "Bobby" appears at morning assembly and waits for them as they returned, "jumping up and down and barking in delight. For him, there was no doubt that we were men."[23] Is Bobby's visible enjoyment, located in and expressed through the body, an address to the human-animal? Is speech to be located as an a priori function of a species (biological property) or in the acknowledgment of a non-relation which, nonetheless, demands a response? *Is the animal that enjoys the animal that speaks?* And what does it mean to respond to the enjoying/speaking animal?

Ethics, for Levinas, passes through the encounter with the face of the other man, that is, the encounter with the humanism of the other. It is a phenomenological encounter. Does the animal have a face? Does the dog Bobby, for instance? In his first Seminar, Lacan had already emphasised the importance of the visual apparatus in human and other-than-human animals in establishing the register of the imaginary. Levinas's reluctance to address the question of the animal philosophically is understandable given the prominence that the face occupies in his thought. In an interview given in 1976, he claims that "one cannot entirely refuse the face of an animal" and more phenomenological analysis is needed: "I don't know if a snake has a face. I can't answer that question."[24] But after stating that "the ethical extends to all living beings," he adds: "But the prototype of this is human ethics.... The animal suffers. It is because we, as human, know what suffering is that we can have this obligation."[25] Yet, Levinas's brief phenomenological description of this encounter reveals the limitations encountered by philosophical discourse in addressing these questions. Bobby is not the heir of the humanist discourse that was open to Argo, Ulysses/Odysseus's dog, for the camp is not Ithaca but a "nowhere." Levinas adds: "This dog was the last Kantian in Nazi Germany, without the brain needed to universalize maxims and drives. He was a descendant of the dogs of Egypt. And his friendly growling, his animal faith, was born from the silence of his forefathers on the bank of the Nile."[26] There is an echo of the Real in the growling of Levinas's dogs—this animal sound which signifies but is not fully inserted in the chain of signifiers of the symbolic order, and yet it is knotted to the symbolic in the animal's response to the event as well as the animal's demand for recognition. The universalisation of laws, principles, categories of thought, which in Lacan's work constitute the symbolic and the work of the symbolic in the drive, cannot account for this encounter since the symbolic cannot fully become itself: a remainder is produced in the symbolic operation, a remainder *extimate* to language (the real of *jouissance*). Met by the limits of his own discourse, the philosopher has to make recourse to the metaphorical and the poetic in order to articulate what remains inaccessible to human language and, therefore, what remains outside of signification. Animal speech is a growling, which does not carry beliefs but animal faith (that central element of social bond appropriated by human *culture*), one that speaks as the unsignifiable but also one that speaks through millennial history. There is no doubt that Levinas's articulation of this memory is steeped into anthropomorphism. But a close reading of this piece reveals much more, perhaps unbeknownst to the philosopher himself. Levinas needs to make use of the poetic in order to come to terms with this *event*, the encounter with Bobby who, in his joyous barking, restitutes to the prisoners the discarded remainders of their own humanity: love as the gift of one's lack. It is a moment of suspension that seems to stage what Lacan calls the non-relation of desire but also, perhaps uncannily, the *lalangue* subtending the communication of the animal.[27] Levinas's elegiac tone speaks to the ambivalence of this encounter, which is supposed to give a shred of joy in the horror of the camp but ends up confronting the prisoners with their lack—for the gift of love comes from the animal, not the human. The animal gives to the camp prisoner what the dog does not have (humanity) and what the (human) prisoner does not want (a humanity inconsistent with anthropocentric humanism). In this non-relation it is difficult not be reminded of Lacan's statement in *Seminar XII* that "love is to give what one does not have to someone who does not want it."[28] In thinking back to this event, the philosopher cannot speak about the animal desire. Yet he strongly suspects that terms like culture, history, and speech can hardly be reserved for the human only. What right, in the end, do we have in categorizing "the animal" as such? Resorting to poetic speech, and refraining from explanation and argumentation, the philosopher reminds us that although we are at a loss in addressing the question of the animal, we have a duty to respond to the possibility of an address.

Desire-In-The-World

How to differentiate this understanding of the animal from the anthropomorphic projections that have informed the development of Western thought? This question becomes more urgent after Darwin's demonstration of the continuity—and kinship—of the human and other-than-human animal worlds but it remains a pertinent one and lingers, barely undetected, in Lacan's discourse too.[29] It is useful to remember Georges Bataille's thinking of the "poetic fallacy of animality": "Nothing, as a matter of fact, is more closed to us than this animal life from which we are descended" and into which only poetry can perhaps in slipping towards the unknowable.[30] Yet, for Bataille, the animal world is "immanence and immediacy," for such an animal world, "which is closed to us, is so to the extent that we cannot discern in it an ability to transcend itself."[31] Bataille remains uncertain about such immanence, in which we seem to discern embryonic possibilities, but not the absolute establishment of the ability to transcend itself. But such reflections can only be borne by the human mind. When we look into animality, for Bataille, we see nothing, for the animal moves in the world like "water in water" and cannot be known .[32] As Jill Marsden observes, the world of animality is unfathomable and unknowable: only from the point of view of consciousness is the animal "lacking."[33] Bataille's frequent discussions of animality in *Theory of Religion* would require a more detailed analysis than I can include here. But it is worth highlighting what also remarked by Marsden: "This animality, of which we are a part yet from which we strive to distinguish ourselves, hovers at the edge of consciousness, extending its glimmer into the night of unknowing."[34] A glimmer of the truth, then, rather than the knowledge of the *savoir-faire*.

The same uncertainty about the substantial otherness of the animal appears in Martin Heidegger's thought. In his inquiry into the *being-in-the-world* that constitutes *Dasein*, which he will conclude to be belonging uniquely to the human, Heidegger comes back to the question of animality at several points in his work, always with different inflections and (non)explanations. One cannot but wonder whether this insistence betrays an anxiety in the philosopher regarding the discrete differentiation between the human and the other-than-human animal. In the formulation of his three thesis—"1) the stone (material object) is *worldless*; 2) the animal is *poor in the world*; 3) man is *world-forming*"—Heidegger describes at length the lizard's basking on a rock in the sun before concluding, without much explanation, that unlike the human, the lizard has its own relation to the rock and the sun; it is not simply "present-at-hand" like an object; yet sun and rock are not given to the lizard in themselves, *as* sun and *as* rock.[35] It is the "as such" (*an sich*) that the animal lacks. Although he admits to the difficulty of establishing a distinction between human and animal relation to the world, Heidegger does not clarify, as Calarco points out, how categories of animal and human are determined beforehand in order to ask this very question nor how a lizard (or the bee which he describes at a different point) could summarise and embody the whole of the "animal" world. In his several attempts to determine animals' different modes of being-in-the-world—a world that humans cannot explore from within—Heidegger makes use of the biological and zoological research of his own time and, in particular, the work of Jakob von Uexküll.[36] Yet Heidegger does not pursue the implications of this material which, despite its partial findings, could open up to different avenues of inquiry. Von Uexküll's notions of *Umwelt* and *Innenwelt*, which in the ethologist is meant to displace the privileged point of view of the human and the moral high ground of the Kantian "I," are used by Heidegger to formulate the idea that animals remain encircled in their environmental world—a unique animal-world to which they have access as a living being, but "which is incapable of further expansion or contraction."[37, 38] In *Seminar I*, Lacan was to take a different approach. In discussing the function of the imaginary in the formation of the Ego-ideal and the ideal ego, he delves into a discussion of the possibility of the imaginary in the animal:

> Let's start with the animal, an animal which is also an ideal, that is to say successful—the unsuccessful one is the animal we managed to capture. This ideal animal gives us a vision of completeness, of fulfilment, because it presupposes the perfect fit, indeed the identity of the *Innenwelt* and the *Umwelt*. That's what makes this living form seductive, as its appearance harmoniously unfolds.[39]

The animal, therefore, is constructed by Man as an ideal on the basis of a presupposed coincidence of *Innenwelt* and *Umwelt*—a coincidence that the human "sick animal" or the other-than-human captive animal have lost. Although the whole section is directed at a discussion of the imaginary, the piece is subtended with an unmistakable critique of philosophical discourse. But before addressing the formation of the imaginary in humans, Lacan subverts this discourse by remarking the enormous importance of the image of the body for the animal. He notes that it is enough the appearance of a phenotype to release sexual behavior, which points to the significance of the imaginary phenomenon: "The animal makes a real object coincide with the image within him. And, what is more, as is indicated in Freud's texts, I would add that the coincidence of the image with a real object strengthens it, gives it substance, embodiment."[40]

Which are the animals that Heidegger discusses to substantiate his theses? As Jacques Derrida notes, Heidegger does not stray from the philosophical tradition, despite his willingness to bring to a close the age of metaphysics—a tradition which collapses the radical diversity of other-than-human animals into one word. The animal, Derrida notes, is a word, an *animot* ("ani-word"), with which humans have given themselves the prerogative to name living beings to which they claim not to belong—an idea first introduced by Nietzsche.[41] The bee, the lizard, the dog—all kinds of animals are reduced to one category, thus revealing the anthropological machine at work. Heidegger's insistence in foreclosing the animal from Dasein becomes even more evident in his attending to the case of domestic animals, in particular the dog. Dogs are not part of the house as a roof would be, he claims, "they '*live*' *with us*. But we don't live with them if living means: *being* in an animal kind of way," and concludes that "the dog does not have a Dasein, it does not exist, it merely lives."[42] In a sudden final reverse shortly after, he notes how the dog feeds with us (but does not really eat), eats with us, is along with us … "and yet not."[43]

Appropriately, Derrida calls this a moment of negation-disavowal. But what does this moment show us? Whose dog is Heidegger talking about? Or to which dog is Heidegger talking? A close-reading of this moment would require more attention than I am able to give here. Yet what becomes apparent in Heidegger's language is a tension between being and non-being, existence and *merely* living (which would merit a whole different discussion in relation to biopolitics), the being-present to oneself and the drives and desires that belong to and encircle the animal world.[44] In this specular game of reflection, we have the impression that Heidegger himself has remained trapped and is not even sure he can (or wants) to get out. We have reached the impasse of language as the house of being, or the prison-house of language, which encircles the human and enraptures us in its imaginary dimension as well as in its symbolic dimension. We cannot *be* outside our language. It is no wonder that at the end of his life Heidegger turned towards poetic language as a way to allow the Truth of the Real to emerge. It is also no wonder that Heidegger's thought is turned inside out in Lacan's work.

On the one hand, we could respond to Heidegger on the basis of the important advancements in scientific knowledge in the fields of ethology, cognition, and animal communication. Donald Griffin and Gayle Speck, for example, have discussed experimental and observational data showing that many animals (including invertebrates) experience some form of subjective consciousness and awareness, which is expressed through use of gesture, episodic and long-term memory, explicit learning, communication and report of one's subjective experience, as well as knowledge of one's knowing, or meta-knowledge.[45] Antonio Preti has reviewed the data on animal suicide or self-injury in animals subject to sudden or restrictive confinement, pets separated from their owners, or conditions of social deprivation.[46] Such data is at odds with the notion that animals do not have access to the "as such," which raises the question of the reasons for insisting on such assumed dividing line (the foreclosure of scientific discourse).[47] The elements are there to speak of the unconscious in the animal but also of the social unconscious of which animals are part. Ethological research also played a role in Lacan's own thinking, as Buse and Hub Zwart have shown, but resorting to it *to explain* the work of the Unconscious would prove unfruitful: it would amount to confusing Knowledge with Truth. Bataille's reflections on unknowability reveal indeed, in poetic language, not only the impossibility of knowledge but also the epistemic violence that knowledge produces (that is, the mastery that drives it). On the other hand, we could respond to Heidegger with the myriad of anecdotes from everyday experience in our unique relation to equally unique animals—domestic or otherwise—showing that being-in-the-world partakes more of being-with than accessibility to the "as such" (as in the case of humans). This is perhaps the important lesson of Lacan's psychoanalysis: the decentring of being and thought and the void that constitutes our (non)being. The Lacanian subject names an impossibility, a cut, in the symbolic: The Real at the very core of its subjectivity, its *extimate* centre. But the claim that such a void (the lack) is unique to the human experience would mean falling, once again, into an anthropocentric perspective that attempts to re-imagine "the animal" as the bearer of the fullness of being because at one with its own "nature."[48]

While maintaining a differentiation between human and animal on the basis of language, Lacan consistently collapses such divisions and opts for a topological movement between *Innenwelt* and *Umwelt* that concerns both animals and humans. Two statements from *Television* are to the point. In the first, before referring to humans as "homme-sick animals" and therefore "domestics (*d'hommestiques*)," he states that "There is no unconscious except for the speaking being. The others, who possess being only through being named—*even if they impose themselves from within the real*—have instinct, namely the knowledge needed for survival. *Yet this is so only for our thought*, which might be inadequate here."[49] Shortly after, he notes that the world (*Umwelt*) "is merely the fantasy through which thought sustains itself—'*reality*' no doubt but to be understood as the grimace of the real."[50] It is only the unknowability of other-than-human animals that we can address as, fundamentally, the unknowability and radical singularity of every other human animal. In consistently recasting *l'homme* in a non-relation with *l'animal*, Lacan's work seems to suggest a topological structure knotting other-than-human and human subjects. As I write, I am musing beside my cat whose life is unfolding in its final months towards his death. And like Derrida in *The Animal That Therefore I Am*, I cannot avoid asking myself: What does my cat want? Who is it that I am looking at when I look into my cat's eyes?

Notes

1. See Robert S. Emmett and David E. Nye, eds., *The Environmental Humanities: A Critical Introduction* (Cambridge, MA: MIT Press, 2017); Hannes Bergthaller and Peter Mortensen, eds., *Framing the Environmental Humanities* (Leiden and Boston: Brill Rodopi, 2018); and Sverker Sörlin, "Environmental Humanities: Why Should Biologists Interested in the Environment Take the Humanities Seriously?" *BioScience* 62, no. 9 (September 2012): 788-89, 10.1525/bio.2012.62.9.2; as well as the journals *Environmental Humanities* and *Resilience: A Journal of the Environmental Humanities*. Ecocritical thought and the *greening* of cultural studies, with the work of scholars such as Lawrence Buell, Rob Nixon, and Ursula K. Heise, were central to the development of the Environmental Humanities.
2. Peter Buse, 'The Dog and the Parakeet: Lacan among the Animals,' *Angelaki: Journal of the Theoretical Humanities* 22, no. 4 (2017): 133-45, 10.1080/0969725X.2017.1406052.
3. Although in his last seminars Lacan adopts a position of psychoanalysis as anti-philosophy, such position should be understood as a continuing interrogation and dialogue with philosophy.
4. Jacques Lacan, *The Knowledge of the Psychoanalyst: Seven Talks at Sainte-Anne, 1971-1972*, trans. Cormac Gallagher, http://www.lacaninireland.com/web/wp-content/uploads/2010/06/Book-19a-The-Knowledge-of-the-Psychoanalyst-1971-1972.pdf (accessed May 30, 2020). Gallagher's translation is from unedited French manuscripts. Lacan gave these talks in the same period that he was holding *Seminar XIX: … Ou pire / … Or worse, 1971-1972*, trans. Cormac Gallagher, http://www.lacaninireland.com.
5. Lacan, *Knowledge of the Psychoanalyst*, II 4 (December 2, 1971); emphasis mine: the quotation also emphasises the failure of reaching being as the structural failure of language.
6. Lacan, *Knowledge of the Psychoanalyst*, I 8 (November 4, 1971).
7. Aristotle uses the expression at several points of his *Ethics* and *Politics*. See *Ethics: Writings from the Complete Works of Aristotle*, ed. Jonathan Barnes and Anthony Kenny, trans. Anthony Kenny (Princeton, NJ: Princeton University Press, 2014); and *Politics: Writings from the Complete Works of Aristotle*, ed. Jonathan Barnes, trans. Benjamin Jowett (Princeton, NJ: Princeton University Press, 2016).

8. René Descartes's formulation is included in the *Principles of Philosophy*, whereas it appears as "Je pense, donc je suis" in the *Discourse on Method*. It is important to point out that Descartes never advocated animal cruelty nor did he claim that animals do not suffer pain. But such pain, for Descartes, cannot be fully experienced as such because the animal lacks language and awareness. Despite this subtle differentiation, the implications of Cartesian thought for the lives of animals remain disastrous. See *A Discourse on Method* (1637), trans. John Veitch (N.p.: N.p.) Gutenberg Project, 2008, https://www.gutenberg.org/files/59/59-h/59-h.htm; and *Selections from the Principles of Philosophy* (Latin ed., 1644), trans. John Veitch (N.p.: N.p.) Gutenberg Project, 2003, http://www.gutenberg.org/cache/epub/4391/pg4391-images.html.
9. Immanuel Kant, "Duties to Animals and Spirits," in *Lectures on Ethics* (1775-1780), trans. Louis Infield (New York: Harper and Row, 1963), 241.
10. Georg Wilhelm Friedrich Hegel, *Philosophy of Nature*, trans. A. V. Miller (Oxford: Oxford University Press, 2004); and Hegel, *The Human Spirit: A Translation of the Jena Lectures on the Philosophy of Spirit* (1805-1806), trans. Leo Rauch (Detroit: Wayne State University Press, 1983).
11. Georg Wilhelm Friedrich Hegel, *Hegel's Phenomenology of Spirit* (1807), trans. A. V. Miller (Oxford: Oxford University Press, 1977).
12. Friedrich Nietzsche, *On the Genealogy of Morals* (1887), trans. Douglas Smith (Oxford: Oxford University Press, 2009).
13. Michel de Montaigne, "An Apology of Raymond Sebond" (ca. 1588), in *The Complete Essays of Montaigne*, trans. Donald Frame (Stanford, CA: Stanford University Press, 1965), 331. The citation is from the section "Man is no better than Animals."
14. David Hume, "Of the Reason of Animals," in *A Treatise of Human Nature* (1739), vol. 1, bk. 1, pt. 3, section xvi (Oxford: Clarendon Press, 1896), 176.
15. Jeremy Bentham, *An Introduction to the Principles of Morals and Legislation*, chap. 17, sect. 1 (London: T. Payne, 1789), cccix; emphasis in the original. Extensive footnote in the section on the limitations of jurisprudence in which Bentham discusses the reduction of animals to the status of things following the neglection of legislation.
16. Jean François Lyotard, *The Differend: Phrases in Dispute*, trans. Georges Van Den Abbeele (Minneapolis: University of Minnesota Press, 1988), 56-57. I use Lyotard's notion of the *différend* to underline the silence that

surrounds Montaigne's and Bentham's own interrogation of the discourse of philosophy—knowledge constituted through the act of silencing animal subjectivity and through the (human) horror imposed on animality.

17. Jacques Lacan, *The Seminar of Jacques Lacan, Book XVII, The Other Side of Psychoanalysis*, ed. Jacques-Alain Miller, trans. Russell Grigg (New York: W. W. Norton, 2007).
18. Giorgio Agamben, *The Open: Man and Animal*, trans. Kevin Attell (Stanford, CA: Stanford University Press, 2004), 33.
19. Emmanuel Levinas, *Ethics and Infinity: Conversations with Philippe Nemo*, trans. Richard A. Cohen (Pittsburgh, PA: Duquesne University Press, 1985).
20. Charles Darwin, "Comparison of the Mental Powers of Man and the Lower Animals," in *The Descent of Man, and Selection in Relation to Sex* (London: John Murray, 1871), vol. 1, chap. 2, 34-69, at 40, http://darwin-online.org.uk/content/frameset?itemID=F939.1&viewtype=image&pageseq=1 (accessed May 30, 2020).
21. Eduardo Kohn, *How Forests Think: Toward an Anthropology beyond the Human* (Berkeley: University of California Press, 2013), 8 and 1.

    In his anthropological work in the Andes, Kohn recounts how human encounters with jaguars are based on the force of "seeing" (1) the opponent as predator or meat. Kohn also cites Terrence W. Deacon's *Incomplete Nature: How Mind Emerged from Matter* (New York: W. W. Norton, 2012), in noting that, as symbolic beings, humans share "other semiotic modalities with the rest of nonhuman biological life" (8).
22. Emmanuel Levinas, "The Name of a Dog, Or, Natural Rights," in *Animal Philosophy: Essential Readings in Continental Thought*, ed. Matthew Calarco and Peter Atterton (London and New York: Continuum, 2004), 47-50. This short article first appeared as "Nom d'un chien ou le droit naturel," in *Difficile Liberté* (Paris: Le livre de poche, 1963), 233. The English translation was included in *Difficult Freedoms*, trans. Seán Hand (Baltimore: Johns Hopkins University Press, 1990), 151-53.
23. Levinas, "Name of a Dog," 48-49.
24. Levinas, "Name of a Dog," 49.
25. Levinas, "Name of a Dog," 50.
26. Levinas, "Name of a Dog," 49.

27. Jacques Lacan, *The Seminar of Jacques Lacan, Book XX: On Feminine Sexuality, The Limits of Love and Knowledge, Encore, 1972-1973*, ed. Jacques-Alain Miller, trans. Bruce Fink (New York and London: W. W. Norton, 1999). The first apparition of the term *lalangue* as language of the Real occurs, appropriately, on the occasion of a lapsus on the part of Lacan with regard to a comment on the *Vocabulaire de la psychanalyse*, which he terms by mistake *Vocabulaire de la philosophie*. In *L'étourdit* (dated July 14, 1972), and in the Seminar *Encore* (June 26, 1973), Lacan will develop the relation of *lalangue* with the language of the maternal and the infant's first babbling. See Dominique Simonney, "Lalangue en questions," *Essaim* 29, no. 2 (2012): 7-16, 10.3917/ess.029.0007.
28. Jacques Lacan, *The Seminar of Jacques Lacan, Book XII: Crucial Problems for Psychoanalysis, 1964-1965*, trans. Cormac Gallagher (March 17, 1965), 165, http://www.lacaninireland.com/web/wp-content/uploads/2010/06/12-Crucial-problems-for-psychoanalysis.pdf (accessed May 30, 2020).
29. For a discussion of the metaphors of animality used by Lacan throughout his work and the significance of their increased use in the late work on the Real, see Buse, "Dog and the Parakeet."
30. Georges Bataille, *Theory of Religion* (1948/1973), trans. Robert Hurley (New York: Zone Books, 1989), 20. The original French, *Théorie de la religion*, was drafted in 1948 and first published by Gallimard in 1973. Throughout his Seminars, Lacan engaged at many points with the work of Martin Heidegger and Georges Bataille.
31. Bataille, *Theory of Religion*, 23.
32. Bataille, *Theory of Religion*, 24.
33. Jill Marsden, "Bataille and the Poetic Fallacy of Animality," in *Animal Philosophy: Essential Readings in Continental Thought*, ed. Matthew Calarco and Peter Atterton (London and New York: Continuum, 2004), 41.
34. Marsden, "Bataille," 41.
35. Martin Heidegger, *The Fundamental Concepts of Metaphysics: World, Finitude, Solitude*, trans. William McNeill and Nicholas Walker (Bloomington: Indiana University Press, 1995), 198. Heidegger's lectures were given in 1929-1930.

36. In *The Open* (Heidegger, *Fundamental Concepts*), Agamben also mentions the work of Hans Driesch, Karl von Baer, and Johannes Müller, of whom Jakob von Uexküll was a student.
37. Heidegger, *Fundamental Concepts*, 198.
38. These terms are used in the original German. *Innenwelt* signifies interiority, literally the interior world. *Umwelt*, instead, is not simply the exterior world. As Agamben points out, Uexküll carefully distinguishes "the *Umgebung*, the objective space in which we see a living being moving, from the *Umwelt*, the environment-world that is constituted by a more or less broad series of elements that he calls "carriers of significance" (*Bedeutungsträger*) or of 'marks' (*Merkmalträger*), which are the only things that interest the animal" (Agamben, *The Open*, p. 40).
39. Jacques Lacan, *The Seminar of Jacques Lacan, Book I: Freud's Papers on Technique, 1953-1954*, ed. Jacques-Alain Miller, trans. John Forrester (New York and London: W. W. Norton, 1991), 137.
40. Lacan, *Seminar of Jacques Lacan, Book I*, 138.
41. Jacques Derrida, *The Animal That Therefore I Am*, ed. Marie-Louise Mallet, trans. David Wills (New York: Fordham University Press, 2008). In a deconstructive gesture, Derrida coins the term *animot* by playing on the assonance of the word, which contains the term *word* (*mot*), with the French plural for *animals* (*animaux*).
42. Heidegger, *Fundamental Concepts*, 210; emphasis in the original.
43. Heidegger, *Fundamental Concepts*, 210.
44. For a discussion of biopolitics in relation to mere life, see Giorgio Agamben, *Homo Sacer: Sovereign Power and Bare Life*, trans. Daniel Heller-Roazen (Stanford, CA: Stanford University Press, 1998).
45. Donald R. Griffin and Gayle B. Speck, "New Evidence of Animal Consciousness," *Animal Cognition* 7 (2004): 5-18, 10.1007/s10071-003-0203-x. See also Stephen Walker's *Animal Thought* (London and New York: Routledge, 1986), for a documentation of the neurophysiological similarity between human and non-human animals. Generally speaking, the presence of pain and pleasure centres across species, and the similarity of the neural mechanism responsible for brain behaviour in vertebrates, have been well documented.
46. Preti, Antonio. "Suicide among Animals: A Review of Evidence." *Psychological Reports* 101, no. 3 (2007): 831-48. 10.2466/2Fpr0.101.3.831-848.

47. Indeed, Bernard E. Rollin has shown how the attribution of "mental states" is dependent upon the possibility of morality and how the exclusion by science of much data for consideration "is in large part determined by our metaphysical commitment and associated values." See Rollin, *The Unheeded Cry: Animal Consciousness, Animal Pain, and Science*, 2nd ed. (Columbia: University of Missouri Press, 2017).
48. In addressing Lacan's position with regard to animality, Hub Zwart reads Lacan as claiming that animals lack the lack. Zwart's article contains a productive discussion of the implications of Lacanian analysis for the realm of ethics—our moral obligations to an ethical treatment of animals—but his claim with regard to the position of the lack in Lacan's discourse is not substantiated by a reading of either *Écrits* or *Seminars*. Admittedly, Lacan seems to gesture in this direction but only in the unravelling of his own reflections on the unconscious. We should also remember the oral delivery underpinning Lacan's Seminars, in typical European teaching style. Although Lacan did have notes, he also produced much of his thought in the course of his delivery, which confines many of the comments on animality to by-examples less thought out than they would otherwise be. As Buse has noted in "Dog and the Parakeet," the status of the animal world undergoes a change as Lacan's thinking moves increasingly toward the register of the Real in his late Seminars. See Hub Zwart, "The Elephant, the Mirror and the Ark: Rereading Lacan's Animal Philosophy in an Era of Ontological Violence and Mass Extinction," *Journal of Critical Animal Studies* 12, no. 1 (2015): 1-32.
49. Jacques Lacan, *Television*, trans. Denis Hollier, Rosalind Krauss, and Annette Michelson. Special issue, ed. Joan Copjec, *October* 40 (Spring 1987): 9; emphasis mine. https://www.jstor.org/stable/i231783.
50. Lacan, *Television*, 10.

Bibliography

Agamben, Giorgio. 1998. *Homo Sacer: Sovereign Power and Bare Life*. Trans. Daniel Heller-Roazen. Stanford, CA: Stanford University Press.

———. 2004. *The Open: Man and Animal*. Trans. Kevin Attell. Stanford, CA: Stanford University Press.

Aristotle. 2014. *Ethics: Writings from the Complete Works of Aristotle*. Ed. Jonathan Barnes and Anthony Kenny & Trans. Anthony Kenny. Princeton, NJ: Princeton University Press.

———. 2016. *Politics: Writings from the Complete Works of Aristotle*. Ed. Jonathan Barnes and Trans. Benjamin Jowett. Princeton, NJ: Princeton University Press.

Bataille, Georges. 1989. *Theory of Religion* (1948/1973). Trans. Robert Hurley. New York: Zone Books.

Bentham, Jeremy. 1789. *An Introduction to the Principles of Morals and Legislation*. London: T. Payne.

Bergthaller, Hannes, and Peter Mortensen, eds. 2018. *Framing the Environmental Humanities*. Leiden and Boston: Brill Rodopi.

Buse, Peter. 2017. The Dog and the Parakeet: Lacan Among the Animals. *Angelaki: Journal of the Theoretical Humanities* 22 (4): 133–145. https://doi.org/10.1080/0969725X.2017.1406052.

Calarco, Matthew. 2004. Heidegger's Zoontology. In *Animal Philosophy: Essential Readings in Continental Thought*, ed. Matthew Calarco and Peter Atterton, 18–30. London and New York: Continuum.

Darwin, Charles. 1871. *The Descent of Man, and Selection in Relation to Sex*. 2 vols. London: John Murray. http://darwin-online.org.uk/content/frameset?itemID=F939.1&viewtype=image&pageseq=1. Accessed 30 May 2020.

Deacon, Terrence W. 2012. *Incomplete Nature: How Mind Emerged from Matter*. New York: W. W. Norton.

Derrida, Jacques. 2008. *The Animal That Therefore I Am*. Ed. Marie-Louise Mallet. Trans. David Wills. New York: Fordham University Press.

Descartes, René. 2003. *Selections from the Principles of Philosophy* (Latin ed., 1644). Trans. John Veitch. N.p.: N.p. Gutenberg Project. http://www.gutenberg.org/cache/epub/4391/pg4391-images.html.

———. 2008. *A Discourse on Method* (1637). Trans. John Veitch. N.p.: N.p. Gutenberg Project. https://www.gutenberg.org/files/59/59-h/59-h.htm.

Emmett, Robert S., and David E. Nye, eds. 2017. *The Environmental Humanities: A Critical Introduction*. Cambridge, MA: MIT Press.

Griffin, Donald R., and Gayle B. Speck. 2004. New Evidence of Animal Consciousness. *Animal Cognition* 7: 5–18. https://doi.org/10.1007/s10071-003-0203-x.

Hegel, Georg Wilhelm Friedrich. 1977. *Hegel's Phenomenology of Spirit* (1807). Trans. A. V. [Arnold Vincent] Miller. Oxford: Oxford University Press.

———. 1983. *Hegel and the Human Spirit: A Translation of the Jena Lectures on the Philosophy of Spirit* (1805-1806). Trans. Leo Rauch. Detroit: Wayne State University Press.

———. 2004. *Hegel's Philosophy of Nature: Part II of the Encyclopaedia of the Philosophical Sciences* (1830). Trans. A. V. [Arnold Vincent] Miller. Oxford: Oxford University Press.

Heidegger, Martin. 1995. *The Fundamental Concepts of Metaphysics: World, Finitude, Solitude* (1929-1930). Trans. William McNeill and Nicholas Walker. Bloomington: Indiana University Press.

Hume, David. 1896. Of the Reason of Animals. In *A Treatise of Human Nature* (1739). Edited by L. A. Selby-Bigge. Volume 1 of 3 vols.; Section 16. Oxford: Clarendon Press.

Kant, Immanuel. 1963. Duties to Animals and Spirits. In *Lectures on Ethics* (1775-1780). Trans. Louis Infield, 239-241. New York: Harper and Row.

Kohn, Eduardo. 2013. *How Forests Think: Toward an Anthropology beyond the Human*. Berkeley: University of California Press.

Lacan, Jacques. 1971. *The Knowledge of the Psychoanalyst: Seven Talks at Sainte-Anne, 1971-1972*. Translated from unedited manuscripts by Cormac Gallagher. http://www.lacaninireland.com/web/wp-content/uploads/2010/06/Book-19a-The-Knowledge-of-the-Psychoanalyst-1971-1972.pdf. Accessed 30 May 2020.

———. 1987. "Television." Trans. Denis Hollier, Rosalind Krauss, and Annette Michelson. Special issue. Ed. Joan Copjec. *October* 40 (Spring): 6-50. https://www.jstor.org/stable/i231783.

———. 1991. *The Seminar of Jacques Lacan, Book I: Freud's Papers on Technique, 1953-1954*. Ed. Jacques-Alain Miller and Trans. John Forrester. New York and London: W. W. Norton.

———. 1999. *The Seminar of Jacques Lacan, Book XX: On Feminine Sexuality, The Limits of Love and Knowledge, Encore, 1972-1973*. Ed. Jacques-Alain Miller and Trans. Bruce Fink. New York and London: W. W. Norton.

———. 2007a. Kant with Sade. In *Écrits: The Complete Edition in English*, Trans. Bruce Fink, 645-69. New York and London: W. W. Norton.

———. 2007b. *The Seminar of Jacques Lacan, Book XVII, The Other Side of Psychoanalysis*. Ed. Jacques-Alain Miller and Trans. Russell Grigg. New York and London: W. W. Norton.

———. *The Seminar of Jacques Lacan, Book XII: Crucial Problems for Psychoanalysis, 1964-1965*. Translated by Cormac Gallagher. http://www.lacaninireland.com/web/wp-content/uploads/2010/06/12-Crucial-problems-for-psychoanalysis.pdf. Accessed 30 May 2020.

Levinas, Emmanuel. 1985. *Ethics and Infinity: Conversations with Philippe Nemo*. Trans. Richard A. Cohen. Pittsburgh, PA: Duquesne University Press.
———. 2004. The Name of a Dog, Or, Natural Rights. In *Animal Philosophy: Essential Readings in Continental Thought*, ed. Matthew Calarco and Peter Atterton, 47–50. London and New York: Continuum.
Marsden, Jill. 2004. Bataille and the Poetic Fallacy of Animality. In *Animal Philosophy: Essential Readings in Continental Thought*, ed. Matthew Calarco and Peter Atterton, 37–44. London and New York: Continuum.
Montaigne, Michel de. 1965. An Apology of Raymond Sebond (1580/1595). In *The Complete Essays of Montaigne*, Trans. Donald Frame, 318-457. Stanford, CA: Stanford University Press.
Nietzsche, Friedrich. 2009. *On the Genealogy of Morals* (1887). Trans. Douglas Smith. Oxford: Oxford University Press.
Preti, Antonio. 2007. Suicide Among Animals: A Review of Evidence. *Psychological Reports* 101 (3): 831–848. https://doi.org/10.2466/2Fpr0.101.3.831-848.
Rollin, Bernard E. 2017. *The Unheeded Cry: Animal Consciousness, Animal Pain, and Science*. 2nd ed. Columbia: University of Missouri Press.
Simonney, Dominique. 2012. Lalangue en questions. *Essaim* 29 (2): 7–16. https://doi.org/10.3917/ess.029.0007.
Sörlin, Sverker. 2012. Environmental Humanities: Why Should Biologists Interested in the Environment Take the Humanities Seriously? *BioScience* 62 (9): 788–789. https://doi.org/10.1525/bio.2012.62.9.2.
Voltaire. 2010. "Animals." *Voltaire's Philosophical Dictionary*. 21-23. New York: Carlton House, 1764. Project Gutenberg. http://www.gutenberg.org/files/18569/18569-h/18569-h.htm#Animals.
Walker, Stephen. 1986. *Animal Thought*. London and New York: Routledge.
Zwart, Hub. 2015. The Elephant, the Mirror and the Ark: Rereading Lacan's Animal Philosophy in an Era of Ontological Violence and Mass Extinction. *Journal of Critical Animal Studies* 12, 1: 1-32. https://ssrn.com/abstract=2555230. Accessed 30 May 2020.

10

Aokigahara Forest: An Aesthetic Space of Residual Surplus

Hilda Fernandez-Alvarez

The Aokigahara or Jukai, which in Japanese means "sea of trees," is a vast forest in Japan that extends for about 30 square kilometres of dense vegetation. Formed over centuries by the hardening of lava from an ancient eruption of sacred Mount Fuji, this beautiful woodland of coniferous trees, mammals, insects, butterflies, and birds, is, at the same time, a landscape associated with suicide. Since early 2000, this forest has attracted the concerned attention of suicide prevention initiatives as it has become the second most popular destination for *suicide tourism*, after the Golden Gate Bridge in San Francisco. Between thirty and one hundred dead bodies are found in the forest every year.

In this chapter, I explore the aesthetic elements that constitute the interstitial space created by this forest and the humans who go there to die voluntarily. I propose to read the Jukai as a psychoanalytic paradigm, a *singular* example of a *universal* psycho-ecological relation, that is, a relation of the human with the environment in light of the unconscious. To advance such an argument, I first analyse the Aokigahara as a *particular*

H. Fernandez-Alvarez (✉)
Simon Fraser University, Vancouver, BC, Canada
e-mail: hilda_fernandez_alvarez@sfu.ca

C. Burnham, P. Kingsbury (eds.), *Lacan and the Environment*, The Palgrave Lacan Series, https://doi.org/10.1007/978-3-030-67205-8_10

forest, an aesthetic and phantasmatic object shaped by the discourse of its own history, markedly associated with death. Second, I discuss the forest qua Forest, as a manifestation of the Real, Jacques Lacan's theory of *das Ding*, and the subject's recourse to aesthetic sublimation by Imaginary/Symbolic means. Third, I theorise the psycho-ecological relation within the Jukai through the late capitalist structure that organises the discursive mechanism of *excess*—surplus *jouissance*, and the consequent *object a* as loss. Finally, as a supplement to the reading of this forest in its Imaginary, Symbolic, and Real aspects, I propose the possibility of a discourse of Nature.

Aokigahara Forest and Its Phantasmatic Representations of Suicide

Suicide in the Jukai is an intricate complexity in itself; its etiology is related to the forest, but also to the issue of suicide in Japan at large. The high rate of suicide in Japan—21.4 deaths per 100,000 people, "well above that of other high-income countries (12.7 deaths per 100,000 people)"[1]—has been understood by some as mainly "due to the stigmatization that surrounds mental health issues in Japan,"[2] which prevents accessing professional support. Specific to suicide in the Aokigahara, Yoshitomo Takahashi interviewed people who attempted suicide at this site. Apart from a contagion effect stemming from the spectacle of mass media reports on this forest, or the cultural representations specific to the Jukai, the attempters expressed a desire to beautify their deaths, to find "a sanctuary where suicide was allowed," [3] to attract more sympathy, and allow them to belong to the same group to those who died there. And Ryuunosuke Akutagawa, a Japanese writer who died by suicide, tell us in his last note: "We might think of a contradiction between the love of Nature and the wish for suicide. However, Nature looks even more beautiful exactly because I am at the end of my life.[4]

The Aokigahara forest has been represented in popular culture through a number novels, such as *Kuroi Jukai* (Black Sea of Trees) (1960) and *Nami no Tou* (Tower of Waves) (1960) by Seichō Matsumoto, in which the latter's portrait of a suicide pact between lovers caused a wave of contagion among the youth in the Japan of the 1960s, similar to the effect of Goethe's 1774 novel *The Sorrows of Young Werther* had in Germany. Hollywood has also exploited this forest, most recently in Gus Van Sant's *The Sea of Trees* (2015) and Jason Zada's *The Forest* (2016). Appealing to virtual dark tourism sentiment, a famous vlogger on YouTube exploited the sensationalistic side of the Aokigahara by showing a dead body on camera, for which, once publicly condemned, he offered an empty apology.

The pervasive associations of the Jukai with death and suicide are the result of various historical layers that have contributed to form its enigmatic representations, even some Japanese people "grew up forbidden to talk about the forest and gasped at its mention."[5] The powerful mystique of this landscape constitutes a fantasy, which in psychoanalysis refers to " a montage of the symbolic and imaginary" that blurs the polarity subject-object to frame reality through the weight of a signifier.[6] Some authors argue that Japan, as a field of study, "first and foremost signifies a psychic object constructed by fantasy,"[7] which they see as a combination of unconscious encoding as well as capitalist and national interests.

Some of the elements that constitute the phantasmatic lure of this forest, based on the material reviewed, could be organized as follows:

(1) Historical stereotypes remain present as an exoticisation of Japan, specifically with regard to ritualistic voluntary death forms, such as *sepukku* (self-disembowelment) or *hara-kiri* (belly cutting) from the Edo period, or the Kamikaze pilots of World War II who crashed intentionally on enemy targets. These forms of suicide, based on honour and shame, rather than entitlement and guilt, were associated by Emile Durkheim with a form of altruistic suicide, as opposed to egoistic or anomic forms.[8] These associations prevailed and yet are vastly stereotypical and unfounded as, for example, *sepukku* practices were supposed "to be held by the warrior class: about 6% of the population" and primarily by men.[9]

(2) The associations with *Ubasute* (senicide): a practice in the eighteenth and nineteenth centuries, which consisted in carrying mainly female elderly to this forest to be left there to die alone in times of famine.

(3) The associations with *yūrei* (ghosts of the dead): These spirits are conceived as related to Japanese Shinto practices, which bring about, in similar ways to other non-Western cosmogonies, the interrelation between spirits of the environment and humans, and where "death provides the vehicle by which a person transitions into unity with nature once more."[10]

(4) The landscape, which refers to the physical geography and the way it is represented. Diverse sources describe the Jukai as a thick dark forest where one can easily get lost as the use of compasses fails in some areas as a consequence of its magnetically charged volcanic soil. Walking along the floor of the forest is a beautiful but eerie experience. "The trees tend to be gnarly and twisted. Their branches interlock like witches' fingers, stretching in a tightly knit canopy that blocks both light and sound."[11] or, "The forest itself as it gets into the mountain is punctuated with caves and a lot of times these caves have been likened to wombs, as though the mountain that we are on is harboring or is impregnated with the souls and the bodies of the people who have come to visit."[12]

(5) The aesthetic knowledge of the place created through the senses and the anxiety associated with them, which is not only a matter of individual sensibility, but also is socially and culturally constructed. Particularly noticeable is the presence of a devouring gaze. Lacan has explained the gaze as an object that looks back at the subject from a luminous point from which reflections emerge forming a picture in which the subject is excluded and, consequently, "subjective subsistence seems to get lost, to be absorbed, and to leave the world behind."[13] Two Japanese poets, who worked together in a *Renga* project (a genre of Japanese collaborative poetry) right at the heart of the Jukai, say of this forest:

trees say nothing
but look carefully
they are prepared to eat me up
at any moment[14]

give it your prayer
or despair
or even death
it takes anything
it is hungry
it has to live[15]

This forest presents us with a gothic aesthetic, which has an incline for the romantic, anxiety-ridden, melancholic macabre, and corresponds to the sublime, traditionally defined as an impressive phenomenon of powerful nature that causes terror.

The Aesthetic of the Forest: *das Ding* or Its Sublimation

For a moment, let us pause our reflection on the Jukai to concentrate on the concept of forest qua *universal Forest*, which is an example of Immanuel Kant's dynamical sublime. The Forest's sublime is both an aesthetic—or sensible—judgement, as well as a moral one, distinct from beauty (beauty in the sense of the mathematical sublime, which refers to magnitude and ideas of grandeur). The Forest is a dynamical sublime because it constitutes itself through a sensorial excess, impossible to represent, that leads to fear due to its boundlessness: "If nature is to be judged by us dynamically as sublime, it must be represented as arousing fear."[16] However, for Kant, such fear is counterbalanced by the subject's own reasoning on their ability to endure. In his *The Theory of the Sublime from Longinus to Kant*, Robert Dorian explains that the sublime involves the problem of "complex pleasure (pleasure mixed with pain)."[17] The Kantian dynamic sublime of the forest speaks to psychoanalysis as a form of *jouissance*, akin to the erotic or the mystic experience. Yet, while Kant will see it as a matter of moral heroism, resolved by a reasoning of self-preservation, psychoanalysis will read it as a form of transgression that aims at accessing *das Ding*, a term which Lacan locates within the ethical and aesthetical field.

Where could we locate the concept of the Forest in Lacan as a thinker of modernity? For example, the Forest, for Martin Heidegger, as it is expressed in *Holzwege*, or "Off the Beaten Track," refers to the concept of *Lichtung* (clearing the woodways). It suggests an authentic, down-to-earth relationship with the ordinary and regular life in the dark forests of the Schwarzwald (The Black Forest), as well as the rural wisdom of mute peasants: a sort of mythology.[18] For Georges Bataille, in his own sensibility, the Forest would constitute part of an inner experience. It is a natural exaltation, a "dark incandescence,"[19] an anti-utilitarian concern with no other goal than itself, which constitutes an operation of sovereignty and is akin to the phenomenon of eroticism, which "attains in the end the fusion of subject and object, being as subject non-knowledge, as object the unknown."[20] The Forest, for Lacan, beyond a fleeting engagement with it as a metaphor for theory, would correspond to the sublime, a primordial object that allows us to apprehend beauty and life, which signal presence and pure absence. This plenitude of absence constitutes a partial reencounter of *das Ding*, and belongs to the same nature of the letter, that which inscribes but does not signify.

Being the most aesthetically oriented term in the Lacanian jargon, *das Ding* is relevant for our approach to the Forest because it provides us with the prefigured object that it incites in us. *Das Ding*, both a figuration of a prehistoric and absolute Other that is impossible to forget and that one suppose will find again, constitutes an excluded center that contributes to the subject's formation for various reasons: *das Ding* not only serves as reference for the subject's wishes and expectations, but it also constitutes the first outside, hostile, and strange (*Fremde*), [21] and the first imprint of the good. This good causes a tropism towards the pleasure principle—for Lacan, parallel to the reality principle—which allows an orientation towards life and desire, always disrupted by the excess of *jouissance*. Lacan extracts this term of strong philosophical tradition—the impossible Kantian noumenon, the Hegelian contradiction—from Sigmund Freud's *Entwurf* (Project for a scientific psychology of 1895), and it constitutes an important notion for understanding the Lacanian Real, outside of the Symbolic-Imaginary unconscious.

Akin to the experiences of the sea or the night, the immensity of the Forest creates a consciousness of grandeur that, for Gaston Bachelard in his *The Poetics of Space*, constitutes "the world that bears the mark of infinity,"[22] and which reveals the subject through daydreaming. The subject in front of the forest is then a being of open imagination who has been promoted to "the dignity of the admiring being."[23] Bachelard's view coincides with Lacan's *das Ding* when he states that consciousness of immensity, a primary effect of the forest, originates in "the body of impressions which have little connection to the geographical information (or reality)."[24] The quietness, the sense of aliveness, or the tranquillity that one experiences while hiking in the Forest mirrors an inner peace. As the poet Rainer Maria Rilke reminds us, "if you want to achieve the existence of the tree invest it with inner space.[25]

Had hikers embraced *das Ding* in their forest walk they would have transgressed the good of society, as this good demands the renunciation of the sublime; it is forbidden because it is incompatible with life. The sublime, in Freud, refers to that which the movement of sublimation aspires to and simultaneously limits; the most important vicissitude of the drive, the sublime changes both object and aim and finds satisfaction in "a higher social or ethical valuation."[26] Later, Freud renders sublimation as a salient feature of cultural development technique to fence suffering. Lacan works on his own approach to the sublime in *Seminar VII* and partially rejects the Freudian notion of art as a sublimated good because the value of art works as good occurs within the capitalist discourse, when they become commodities. Instead of predetermining the outcome of a sublimated form, Lacan proposes to define it by the logical mechanism that entails raising an object "to the dignity of the Thing."[27] The Forest constitutes an aesthetic phenomenon of the sublime because such landscape has been raised to the dignity of the Thing, engendering a rapture similar to those evoked by childhood tales of enchanted forests from any folk culture.

What do we know of this radical Other, of such intense immensity constituted by the forest, falling on us as Magritte's thin air in "The Castle of the Pyrenees"? The Thing is located beyond the world represented (*Vortworstellung*) or presented (*sache-vorstellung*), is beyond the signified, exceeding the subject's conscious perception. Yet, the proximity with the Forest as *das Ding*, the impossibility of internalizing its immensity, reaches a threshold when it is time to finish hiking and get back home or to the camp site after a *forest bathing*. Staying longer unwillingly (that is, getting lost) might lead to trauma; staying forever through the suicidal act, a return to the organic state of Nirvana.

Back to the Jukai. The sublime sought after in the ideal of becoming one with this portentous space of nature, the forest, is either sublimated or is fully embraced in a self-effacing act. Transforming the sublime into beauty is part of the pleasure principle's function, grounded in Eros. A poet well exemplifies this sublimation since, tempted by the abyssal transgression of embracing Freud's "oceanic feeling,"[28] he harnessed life in this way:

> Dig a hole in its tender skin of moss
> Touch the heart
> Petrified death underneath
>
> Take its breath
> Hide it inside you
> And go home[29]

Instead, the suicidal individual who stays in this forest forever, renders their life to the sublime through the act of death, perhaps the only way to fully embrace the sublime. This forest might have acted as *a bait* for the unfolding of their death wish. Thus, suicide is clearly not a sublimation as this demands the transformation of the object, one in which lack works as a point of departure and ongoing support.[30] The suicidal subject sublimates nothing because they become the lack itself, incarnating *object a*, a residue or remainder of the Thing.

Reducing the Thing to Produce Excess (Surplus *Jouissance*) and Loss (*Residue*) within Discourse

Thus far, we have been discussing the Jukai's unique phantasmatic reputation and the Forest in its universality as a form of sublimation of *das Ding*. Yet these events do not occur in a void; the experiences of individuals and collectivities are always the result of a discourse that has produced its effects. In this section, we approach those elements through Lacan's theory of discourse. It is not my intention to explain this complex theory here,[31] yet its relevance resides in allowing us to illuminate how the social bond is constituted within a group sharing a positionality with regard to a number of variables such as desire, Other, truth, and the object of surplus *jouissance*. This approach affords to map the topological relation between modes of representation and production. Lacan has defined five basic forms of social link,[32] constituted by "fundamental … stable relations … of one signifier to another" that subsist in language without the need of speech.[33]

About the Aokigahara, a poet states, "In certain places, the layers of words are so close to the surface that we cannot ignore them. They are palpable, alive and tangible."[34] This appears to contradict Lacan's emphasis on how discourse, sustained by language, is without words. We ought to remember that Lacanian discourse involves both the levels of representation (speech, the symbolic, and the imaginary) and production (the embodiment of *jouissance*) which are divergent and antagonistic, yet topologically stand in a simultaneous relation of continuity and rupture.[35] Thus, the relations between humans and the forest happening in the Jukai under late capitalist discourse cannot be solely described by representations; it is necessary to engage with the signifier, in its radical autonomy, and trace its material consequences.

Since discourse always regards the transmission of knowledge—that is to say, what we are able to think—in relation to *jouissance*, we ought to inquire: Can we know of *jouissance* in the Jukai? Although a condition of *jouissance* is to remain obscure, I argue that in the Aokigahara forest there are two deeply connected discursive effects that cause *jouissance* and that are located in the body of the suicidal individual. Such residual body, traumatic in nature, constitutes a paradox: a form of *loss* (the loss of life and its potentialities), as much as an *excess* (the product obtained by the very renunciation of the common good of life).

Object a, the remainder of the Thing, indexes the presence of *jouissance* which is only known by the effects on the body, marked by the signifier. *Object a* is simultaneously the cause of desire and the object of anxiety, which emerges—in its peculiar negative positivity—when the drive, circulating the orifices of the body, approaches desire. Lacan assessed *jouissance* as transgression, though he would later pose *object a* as an excess emanating from discourse, a paradoxical production of surplus *jouissance* created through a lack and the renunciation of pleasure.

Different from Herbert Marcuse's repressive desublimation described in *One-Dimensional Man*, which consist in a massified immediate gratification, sanctioned by the estate for the sake of "extend[ing] liberty while intensifying domination."[36] Lacan's surplus *jouissance*, rather than an exclusive external force of domination, is a pervasive superegoic command to enjoy connected to discourse. We hear the imperative to enjoy, and we (j'oui)ssance. Surplus *jouissance*, *plus du jouir*, which means both more *jouissance* and no more *jouissance*, is what "fabricates object a" within a given discourse.[37] Akin to Freud's *Lustgewinn*, "a gain of pleasure" he observed in wit, this surplus is a subjective response that searches for enjoyment, incrementally, while producing more excess—tension, pleasurable distress, the moment one experiences loss. Surplus *jouissance* is homologous to Karl Marx's surplus value and shares its opacity; it refers not only to the enjoyment without enjoyment, or losing pleasure while searching an imagined supplementary enjoyment, but also the mysterious production of something that does not satisfy as it is constituted by a lack-in-enjoying (*manqué-à-jouir*).[38]

The suicides in the Jukai escalated in 1998 when the Asian Financial Crisis saw the collapse of their currency and the increase of individual private debt. A Marxist reading sees this forest's suicides as the result of specific conditions of the emergence of capitalism in Japan, which maintained "an intimate and unbroken cultural thread running between the old feudal guilds (*za*) and the development of capitalist corporations" that has impacted forms of labour practices in present time, ranging from excess work—*karoshi* (suicide related to overwork)—to the insecurities of the casualization of work linked to digital phenomena such as internet suicide pacts seeking a sense of purpose (*ikigai*).[39] The production of ontological anxieties caused by overworking and social isolation happened to be a structural characteristic of the capitalist discourse, whose key aspect is the production of surplus *jouissance*, characterised by Lacan as both an insatiable accumulation on the side of capital, and an overproduction and unsatisfying excessive consumption for humans. In this reading, some deaths in the Jukai are caused by labour exploitation and alienation, and this is not a metaphor, but a materialized production, an effect of the capitalist discourse.

Another reading of the Jukai's phenomenon connects it to practical reason on the part of the suicidal individual with regard to Japan's real estate industry. The apartments or houses in which people have completed suicide decrease in value as those properties are perceived as uninhabitable because of ghosts; consequently, real estate agents "demand very large sums for loss of income from rent: sometimes for all the other apartments in a building as well as for the performance of purification ceremonies."[40] Here, the person who died by suicide leaves several residues of a loss, not only suffering for their love one's, but also their spirit's spectrality is financially siphoned out by the system, profiting from their deaths.

Could we consider the aesthetic phenomenon of the Jukai as a psycho-ecological relationship between this forest and the suicidal individual to subvert the insatiable consumption and self-valourisation of capital? Suicide might be seen by some as an exit from capitalism, yet the reality of thriving dark tourisms, which profits on guided expeditions for those searching for a fright in the Jukai,[41] prevent us from mistaking suicide as an accomplished act, fully outside of discourse. Lacan said in *Television*, "Suicide is the only act that can succeed without misfiring,"[42] and such a statement can lead to poor outcomes, for example, when a therapist misuses this explosive signifiers irresponsibly when dealing with the suicidal person in treatment, but also locating suicide as "a successful act," fully outside of discourse, might obscure its effects within. Lacan, in his dialectical engagement with paradoxes, also has stated in *Talking to Brick Walls*, that all acts, including suicide, "are failed ones."[43] The paradox of an act resides in the transformation of a subject by erasing it: If the subject dies, then the subject is lost; if the subject is alive, then it cannot *fully* transform itself or it would disappear. I have argued elsewhere[44] that the tragic logic of suicide is structured around desire, anxiety, and *jouissance*, and that this logic constitutes a full separation from the Other, source of signifiers, and yet, suicide deaths, as these prior examples demonstrate, remain assimilated by discourse.

Discourse assimilates residues into the system by hiding them. Typically, environmental residues are defined as pollution by plastic, lead, toxins, radioactive fallout, or litter, affecting water, air, soil, animals, and humans. In the Aokigahara, the remains are constituted by the flesh and bones of the dead individuals and they are ostensibly left on the surface. Those, however, are not the only *tangible* residues of social bonding within capitalism, also their negativised positivity imprints death upon the landscape and on those loved ones left behind. Lucas Pohl proposes to think of psychoanalysis as an object-disoriented ontology, "to follow the transformation of an object from being locatable in reality to being unlocatable in the Real."[45] If one of the crucial tasks of individual psychoanalysis consists in subverting the logic of surplus *jouissance* to account for the *excess* and *loss* in which the subject is implicated, to find its cause—the Master signifier—in discourse; at the socio-political level and with regards to nature, we face a similar task: How do we account for productive excessive *jouissance* in light of the insistent Master's surplus value and surplus *jouissance*?

Towards a Psycho-Ecological Relationship with Nature

The excessive *jouissance* produced by the phenomena of the Jukai, a psycho-ecological relation between a community of trees and dead humans who aspire to a community, has created a sublime, sacred space. In its tragedy, the Aokigahara shows us the effects of death as residues, imprinted distinctly in the diverse environmental scales: At the level of the individual, the inscription of finitude of a form of *jouissance* linked to the subject's lineage; at the level of the landscape, a phantasmatic imprint of culture; and in the Real, the reinserting of their losses into the capitalist discourse.

Contra Lacan's statement that "there is not discourse of nature,"[46] I wonder if this forest, qua Nature, could produce a discourse that facilitates accounting and inscribing hidden residues of loss and gain to modify the brutal reification of both Nature and humanity that we experience in the psycho-ecological relations of our era.

Scientific discourse tells us of a community of trees that communicate with each other through complex semiotic and biochemical patterns created by the mycelium—fungal networks—that share carbon storage not only with their kin, but also with other species, in highly complex ways, including the protection of the most vulnerable or the collaboration against predators, among many other complex exchanges.[47] If a discourse of Nature is possible, it will need to speak partially outside anthropocentric reason, and instead speak through the hybrid assemblage of interactions between the forest and humans, as well as between the forest and other nonhuman entities. If a discourse of Nature is possible, it should bestow the other-than-human with agency and rights, as well as determining it as a common good, designating who is better prepared to guard such rights. If a discourse of Nature is possible, it should account for its waste, an indicator of order and its political limits.[48]

The era of the Anthropocene, defined by capitalist overproduction through exploitative labour, dispossession, and forced consumption, has never accounted for its inherent abstraction and waste. Traumatic residues of the capitalist (dis)order are materialised and absorbed by the most exploited among us,[49] while society is unable to account for the infringements on human dignity and nature's preservation. If human differentiation from nature requires no domination and recognition of mutually co-constituted relationships,[50] and if collectivity defines what constitutes the moral good but has difficulties in negotiating it, then a discourse of Nature would require a location of sacred spaces in nature, a traditional trope in most cosmogonies, to de-reify nature.

The aesthetic space of the Jukai demonstrates that capitalism profits with sacredness, and that the sacrifice inhered in sacredness fell on those who kill themselves. In light of our environmental crisis, we require a discussion within our communities of our place of sacrifice, to restore, according to Bataille's *The Accursed Share*, that which "servile use has degraded, rendered profane,"[51] and to determine what expenditure or destruction we accept and desire. It does not have to be suicide, particularly not the collective suicide that capitalistic governments and global corporations are leading us towards. Perhaps, instead, it is about sacrificing our surpluses—value and *jouissance*—and sacrificing the interests of Capital?

Notes

1. World Health Organization, "Japan Turning a Corner in Suicide Prevention," *World Health Organization*, November 2015, https://www.who.int/mental_health/suicide-prevention/japan_story/en/ (accessed February 14, 2020).
2. Tony McKenna, "The Suicide Forest: A Marxist Analysis of the High Suicide Rate in Japan," *Rethinking Marxism: A Journal of Economics, Culture & Society* 27, no. 2 (2015): 294, https://doi.org/10.1080/08935696.2015.1007789.
3. Yoshitomo Takahashi, "Aokigahara-jukai: Suicide and Amnesia in Mt. Juji's Black Forest," *Suicide and Life-Threatening Behavior* 18, no. 2 (1988): 173, https://doi.org/10.1111/j.1943-278X.1988.tb00150.x.
4. Ryuunosuke Akutagawa in Marc Etkind, *... Or Not to Be: A Collection of Suicide Notes* (New York: Riverhead Books, 1997), 92.
5. Anastasia O'Keefe, "Suicide in Japan," in *Confronting Death: College Students on the Community of Mortals*, ed. Alfred G. Killilea and Dylan D. Lynch (Bloomington, IN: iUniverse, 2013), 205.
6. Sigmund Freud talked about fantasy as a space between subject and object, in "Group Psychology and the Analysis of the Ego" (1921), in *The Standard Edition of the Complete Psychological Works of Sigmund Freud. Volume XVIII: 1920–1922: Beyond the Pleasure Principle, Group Psychology, and Other Works*, ed. and trans. James Strachey et al. (London: Hogarth, 1973), 80. Jacques Lacan utilises the diamond shape to separate and unite subject from object, stating that the "Signifier does not designate what is not there, it engenders it. What is not there, at the origin, is the subject itself," in Jacques Lacan, "Seminar 1: Wednesday, November 16, 1966," in *The Seminar of Jacques Lacan, Book XIV: The Logic of Phantasy, 1966–1967*, trans. Cormac Gallagher, http://www.lacaninireland.com/web/wp-content/uploads/2010/06/14-Logic-of-Phantasy-Complete.pdf (accessed February 14, 2020).
7. Nina Cornyetz and J. Keith Vincent, "Introduction," in *Perversion and Modern Japan: Psychoanalysis, Literature, Culture*, ed. Nina Cornyetz and J. Keith Vincent (New York: Routledge, 2010), 13.
8. Emile Durkheim, *Suicide: A Study in Sociology*, trans. John A. Spaulding and George Simpson (1952; Florence: Routledge, 2002), 175, https://ebookcentral-proquest-com.proxy.lib.sfu.ca/lib/sfu-ebooks/detail.action?docID=242097 (accessed February 14, 2020).

9. Mary Picone, "Suicide and the Afterlife: Popular Religion and the Standardisation of 'Culture' in Japan," *Culture, Medicine, and Psychiatry* 36, no. 2 (2012): 392, https://doi.org/10.1007/s11013-012-9261-3.
10. McKenna, "Suicide Forest," 294.
11. McKenna, "Suicide Forest," 293.
12. *Art of Now*, "Atmosfears," Featuring Jordan Smith, Yasuhiro Yotsumoto, Takako Arai, and Sayaka Osaki, Produced by Anishka Sharma and Barney Savage, on *BBC Radio 4*, aired September 10, 2018, https://www.bbc.co.uk/sounds/play/b0bh431g (accessed February 14, 2020).
13. Jacques Lacan, *The Seminar of Jacques Lacan, Book X: Anxiety, 1962–1963*, ed. Jacques-Alain Miller, trans. A. R. Price (New York: Polity Press, 2014), 241.
14. Sayaka Osak, in *Art of Now*, "Atmosfears."
15. Yasuhiro Yotsumoto, in *Art of Now*, "Atmosfears."
16. Immanuel Kant, *Critique of Practical Judgment*, 5:260, in *The Theory of the Sublime from Longinus to Kant*, by Robert Doran (New York: Cambridge University Press, 2015), 241, https://doi.org/10.1017/CBO9781316182017.
17. Doran, *The Theory of the Sublime*, 241.
18. Martin Heidegger, *Off the Beaten Track*, ed. and trans. Julian Young and Kenneth Haynes (Cambridge: Cambridge University Press, 2002), xx.
19. Georges Bataille, *Essential Writings*, ed. Michael Richardson (London: Sage, 1998), 142.
20. Bataille, *Essential Writings*, 142.
21. In many languages, the etymology of the word "forest" comes from the Latin *foris* (outside) or *forestis* (wood outside), semantically related to the word foreign.
22. Bachelard, *Poetics of Space*, 183.
23. Bachelard, *Poetics of Space*, 184.
24. Bachelard, *Poetics of Space*, 185.
25. Rainer Maria Rilke, "Poem" 1924, quoted in Gaston Bachelard, *The Poetics of Space: The Classic Look at How We Experience Intimate Places*, trans. Maria Jolas (Boston: Beacon Press, 1994), 200.
26. Sigmund Freud, "Two Encyclopaedia Articles" (1923), in *The Standard Edition of the Complete Psychological Works of Sigmund Freud. Volume XVIII: 1920–1922: Beyond the Pleasure Principle, Group Psychology, and Other Works*, ed. and trans. James Strachey et al. (London: Hogarth Press, 1973), 256.

27. Jacques Lacan, *The Seminar of Jacques Lacan, Book VII: The Ethics of Psychoanalysis, 1959–1960*, ed. Jacques-Alain Miller, trans. Dennis Porter (New York: W. W. Norton, 1992), 112.
28. Sigmund Freud, "Civilization and Its Discontent" (1930), in *The Standard Edition of the Complete Psychological Works of Sigmund Freud. Volume XXI, 1927–1931: The Future of an Illusion, Civilization and Its Discontents and Other Works*, ed. and trans. James Strachey et al.(London: Hogarth, 1973), 65.
29. Yotsumoto in *Art of Now*, "Atmosfears." Fragment of the poem.
30. Lacan, *Seminar XIV*, August 3, 1967, in *Book XIV: The Logic of Phantasy*.
31. For a thorough discussion of Lacan's theory of discourses, see Mark Bracher, Marshall W. Alcorn, Jr., Ronald J. Corthell, and Françoise Massardier-Kenney, eds., *Lacanian Theory of Discourse: Subject, Structure, and Society* (New York: New York University Press, 1994); Paul Verhaeghe, *Beyond Gender: From Subject to Drive* (New York: Other Press, 2001); Slavoj Žižek, "Jacques Lacan's Four Discourses," *lacan.com* (2006), http://www.lacan.com/zizfour.htm (accessed February 14, 2020); and Stjin Vanheule, "Capitalist Discourse, Subjectivity and Lacanian Psychoanalysis," *Frontiers in Psychology* 7 (2016): 1–14, https://doi.org/10.3389/fpsyg.2016.01948.
32. The four discourses map structures that host overdetermined embodied practices in algebraic formulas that account for distinct social bonding: University, Master, Hysteric, and Psychoanalysis. Jacques Lacan, *The Seminar of Jacques Lacan, Book XVII: The Other Side of Psychoanalysis*, trans. Russell Grigg (New York: W. W. Norton, 1999).
33. Lacan, *Seminar XVII*, April 19, 1967, in *Book XIV: The Logic of Phantasy*.
34. *Art of Now*, "Atmosfears."
35. Samo Tomšič, *The Capitalist Unconscious: Marx and Lacan* (New York: Verso, 2015), 47.
36. Herbert Marcuse, *One-Dimensional Man: Studies in the Ideology of Advanced Industrial Society* (Boston: Beacon Press, 1964), 72.
37. Jacques Lacan, *Seminar I*, November 13, 1968, in *The Seminar of Jacques Lacan, Book XVI: From an Other to the other, 1968–1969*, trans. Cormac Gallagher, http://www.lacaninireland.com/web/wp-content/uploads/2010/06/Book-16-from-an-Other-to-the-other.pdf (accessed February 14, 2020).
38. Jacques Lacan, *The Seminar of Jacques Lacan, Book XVIII: On a discourse that might not be a semblance, 1971*, trans. Cormac Gallagher, 22, http://

www.lacaninireland.com/web/wp-content/uploads/2010/06/Book-18-On-a-discourse-that-might-not-be-a-semblance.pdf (accessed February 14, 2020).
39. McKenna, "Suicide Forest," 299, 301.
40. Picone, "Suicide and the Afterlife," 397.
41. Picone, "Suicide and the Afterlife," 397.
42. Jacques Lacan, *Television*, trans. Denis Hollier, Rosalind Krauss, and Annette Michelson; *A Challenge to the Psychoanalytic Establishment*, trans. Jeffrey Mehlman, ed. Joan Copjec (New York: Norton, 1990), 43.
43. Jacques Lacan, *Talking to Brick Walls: A Series of Presentations in the Chapel at Sainte-Anne Hospital*, trans. Adrian R. Price (Cambridge: Polity Press, 2017), 25–26.
44. Hilda Fernandez-Alvarez, "Me phynai 'Rather not to be': The Tragic Logic of Suicide between Desire, Anxiety and Jouissance," *Lacunae: APPI International Journal of Lacanian Psychoanalysis* 11 (2015): 28. http://www.academia.edu/download/41052776/H_Fernandez.pdf (accessed June 3, 2020).
45. Lucas Pohl, "Object-Disoriented Geographies: The Ghost Tower of Bangkok and the Topology of Anxiety," *Cultural Geographies* 27, no. 1 (2020): 11 [82], https://doi.org/10.1177/1474474019864984.
46. Jacques Lacan, "Seminar 2: Wednesday, November 20, 1968," in *Book XVI: From an Other to the other*.
47. Diana Beresford-Kroeger, *The Global Forest: Forty Ways the Trees Can Save Us* (New York: Viking, 2010); Ryszard Hunka and Erna Buffie, "What Trees Talk About," *The Nature of Things with David Suzuki*, starring David Suzuki, directed by Ryszard Hunka, *CBC-TV*. November 26, 2017, https://www.cbc.ca/natureofthings/episodes/what-trees-talk-about (accessed June 3, 2020).
48. Lacan, *Seminar VII*, in *Book VII: The Ethics of Psychoanalysis*, 233.
49. Stephan Lessenich states that "the negative external effects of the cost of economic activity [needs] to be 'internalized' somewhere and by someone." Lessenich, *Living Well at Others' Expense: The Hidden Costs of Western Prosperity* (Cambridge: Polity Press, 2019), 30.
50. Theodor Adorno, "Subject and Object," in *Subject and Object: Frankfurt School Writings on Epistemology, Ontology, and Method*, ed. Ruth Groff (New York: Bloomsbury 2014), 153. Adorno proposes that a possible "peace" between subject and object would be only achieved in the "state of distinctness without domination, with the distinct participating in

each other" (153); Eduardo Kohn believes that the forest thinks based on its ability to represent a present and a future based on Pierce's semiosis, for which a sign represents something for someone. Kohn, *How Forests Think: Toward an Anthropology Beyond the Human* (Los Angeles: University of California Press, 2013), 24.

51. Georges Bataille, *The Accursed Share: An Essay on General Economy* (1967; New York: Zone Books, 1988), 55.

Bibliography

Adorno, Theodor. 2014. Subject and Object. In *Subject and Object: Frankfurt School Writings on Epistemology, Ontology, and Method*, ed. Ruth Groff, 149–164. New York: Bloomsbury.

Art of Now. 2018. "Atmosfears," Featuring Jordan Smith, Yasuhiro Yotsumoto, Takako Arai, and Sayaka Osaki, Produced by Anishka Sharma and Barney Savage. *BBC Radio 4*, September 10. Accessed February 14, 2020. https://www.bbc.co.uk/sounds/play/b0bh431g.

Bachelard, Gaston. 1994. *The Poetics of Space: The Classic Look at How We Experience Intimate Places*. Translated by Maria Jolas. Boston: Beacon Press.

Bataille, Georges. 1988. *The Accursed Share: An Essay on General Economy*. New York: Zone Books, 1967.

———. 1998. *Essential Writings*. Edited by Michael Richardson. London: Sage.

Beresford-Kroeger, Diana. 2010. *The Global Forest: Forty Ways the Trees Can Save Us*. New York: Viking.

Bracher, Mark, Marshall W. Alcorn Jr., Ronald J. Corthell, and Françoise Massardier-Kenney, eds. 1994. *Lacanian Theory of Discourse: Subject, Structure, and Society*. New York: New York University Press.

Cornyetz, Nina, and J. Keith Vincent. 2010. Introduction. In *Perversion and Modern Japan: Psychoanalysis, Literature, Culture*, ed. Nina Cornyetz and J. Keith Vincent, 1–20. New York: Routledge.

Doran, Robert. 2015. *The Theory of the Sublime from Longinus to Kant*. New York: Cambridge University Press. https://doi.org/10.1017/CBO9781316182017.

Durkheim, Emile. 2002. *Suicide: A Study in Sociology*. Translated by John A. Spaulding and George Simpson; Florence: Routledge, 1952. Accessed February 14, 2020. https://ebookcentral-proquest-com.proxy.lib.sfu.ca/lib/sfu-ebooks/detail.action?docID=242097.

Etkind, Marc. 1997. *...Or Not to Be: A Collection of Suicide Notes*. New York: Riverhead Books.

Fernandez-Alvarez, Hilda. 2015. Me phynai 'Rather not to be': The Tragic Logic of Suicide between Desire, Anxiety and Jouissance. *Lacunae: APPI International Journal of Lacanian Psychoanalysis* 11: 27–40. Accessed June 3, 2020. http://www.academia.edu/download/41052776/H_Fernandez.pdf.

Freud, Sigmund. 1973a. *The Standard Edition of the Complete Psychological Works of Sigmund Freud. Volume XVIII: 1920–1922: Beyond the Pleasure Principle, Group Psychology, and Other Works*. Edited and translated by James Strachey et al. London: Hogarth.

———. 1973b. *The Standard Edition of the Complete Psychological Works of Sigmund Freud. Volume XXI: 1927–1931: The Future of an Illusion, Civilization and Its Discontents and Other Works*. Edited and translated by James Strachey et al. London: Hogarth.

Heidegger, Martin. 2002. *Off the Beaten Track*. Edited and translated by Julian Young and Kenneth Haynes. Cambridge: Cambridge University Press.

Hunka, Ryszard, and Erna Buffie. 2017. What Trees Talk About. *The Nature of Things with David Suzuki*. Starring David Suzuki. Directed by Ryszard Hunka. *CBC-TV*, November 26. Accessed June 3, 2020. https://www.cbc.ca/natureofthings/episodes/what-trees-talk-about.

Kant, Immanuel. 2015. *Critique of Practical Judgment*, 5:260. In *The Theory of the Sublime from Longinus to Kant*, ed. Robert Doran, 241. New York: Cambridge University Press. https://doi.org/10.1017/CBO9781316182017.

Kohn, Eduardo. 2013. *How Forests Think: Toward an Anthropology Beyond the Human*. Los Angeles: University of California Press.

Lacan, Jacques. 1990. *Television*. Translated by Denis Hollier, Rosalind Krauss, and Annette Michelson; *A Challenge to the Psychoanalytic Establishment*. Translated by Jeffrey Mehlman. Both edited by Joan Copjec. New York: Norton.

———. 1992. *The Seminar of Jacques Lacan, Book VII: The Ethics of Psychoanalysis, 1959–1960*. Edited by Jacques-Alain Miller . Translated by Dennis Porter. New York: W. W. Norton.

———. 1999. *The Seminar of Jacques Lacan, Book XVII: The Other Side of Psychoanalysis*. Translated by Russell Grigg. New York: W. W. Norton.

———. 2014. *The Seminar of Jacques Lacan, Book X: Anxiety, 1962–1963*. Edited by Jacques-Alain Miller. Translated by A. R. Price. New York: Polity Press

———. 2017. *Talking to Brick Walls: A Series of Presentations in the Chapel at Sainte-Anne Hospital*. Translated by Adrian R. Price. Cambridge: Polity Press.

———. *The Seminar of Jacques Lacan, Book XIV: The Logic of Phantasy, 1966–1967*. Translated by Cormac Gallagher. Accessed February 14, 2020. http://www.lacaninireland.com/web/wp-content/uploads/2010/06/14-Logic-of-Phantasy-Complete.pdf.

———. *The Seminar of Jacques Lacan, Book XVI: From an Other to the other, 1968–1969*. Translated by Cormac Gallagher. Accessed February 14, 2020. http://www.lacaninireland.com/web/wp-content/uploads/2010/06/Book-16-from-an-Other-to-the-other.pdf.

———. *The Seminar of Jacques Lacan, Book XVIII: On a discourse that might not be a semblance, 1971*. Translated by Cormac Gallagher. Accessed February 14, 2020. http://www.lacaninireland.com/web/wp-content/uploads/2010/06/Book-18-On-a-discourse-that-might-not-be-a-semblance.pdf.

Lessenich, Stephan. 2019. *Living Well at Others' Expense: The Hidden Costs of Western Prosperity*. Cambridge: Polity Press.

Marcuse, Herbert. 1964. *One-Dimensional Man: Studies in the Ideology of Advanced Industrial Society*. Boston: Beacon Press.

McKenna, Tony. 2015. The Suicide Forest: A Marxist Analysis of the High Suicide Rate in Japan. *Rethinking Marxism: A Journal of Economics, Culture & Society* 27 (2): 293–302. https://doi.org/10.1080/08935696.2015.1007789.

O'Keefe, Anastasia. 2013. Suicide in Japan. In *Confronting Death: College Students on the Community of Mortals*, ed. Alfred G. Killilea and Dylan D. Lynch, 187–209. Bloomington, IN: iUniverse.

Picone, Mary. 2012. Suicide and the Afterlife: Popular Religion and the Standardisation of 'Culture' in Japan. *Culture, Medicine, and Psychiatry* 36 (2): 391–408. https://doi.org/10.1007/s11013-012-9261-3.

Pohl, Lucas. 2020. Object-Disoriented Geographies: The Ghost Tower of Bangkok and the Topology of Anxiety. *Cultural Geographies* 27 (1): 71–84. https://doi.org/10.1177/1474474019864984.

Rilke, Rainer Maria. 1994. "Poem" (1924). In *The Poetics of Space: The Classic Look at How We Experience Intimate Places*. Edited by Gaston Bachelard. Translated by Maria Jolas, 200. Boston: Beacon Press.

Takahashi, Yoshitomo. 1988. Aokigahara-jukai: Suicide and Amnesia in Mt. Juji's Black Forest. *Suicide and Life-Threatening Behavior* 18 (2): 164–175. https://doi.org/10.1111/j.1943-278X.1988.tb00150.x.

Tomšič, Samo. 2015. *The Capitalist Unconscious: Marx and Lacan*. New York: Verso.

Vanheule, Stjin. 2016. Capitalist Discourse, Subjectivity and Lacanian Psychoanalysis. *Frontiers in Psychology* 7: 1–14. https://doi.org/10.3389/fpsyg.2016.01948.

Verhaeghe, Paul. 2001. *Beyond Gender: From Subject to Drive.* New York: Other Press.

World Health Organization. 2015. Japan Turning a Corner in Suicide Prevention, World Health Organization, November. Accessed February 14, 2020. https://www.who.int/mental_health/suicide-prevention/japan_story/en/.

Žižek, Slavoj. 2006. Jacques Lacan's Four Discourses. Accessed February 14, 2020. *lacan.com.* http://www.lacan.com/zizfour.htm.

11

"Some people like…": Misapprehension and Effacement of *Jouissance* in Climate Hostile Advertising

Miguel Rivera

Advertising traffics in fantasy.[1] The premise of advertisement is to draw out the feeling of lack in the viewer and to propose a means to fill that lack.[2] Satisfaction is promised, often through the fantasy of another, turning the viewer into someone else. Commercials often promise self-improvement, not in terms of mental health or physical fitness, but in terms of one's self-conception. The promise of an advertisement is that someone *feels better*, or is *more themselves* when they buy a certain product. What is left, though, is an absence that needs to be filled by consumption. From this solicited lack comes an excess, a compulsion to purchase, one the advertisement itself must disavow. The conceit of the advertisement is often that the viewer is an absolutely rational consumer. Or at the very least, that the purchase of a given product will not give way to what is inevitable: persistent lack. But even as advertisements struggle to titillate consumers through the unconscious, *jouissance* remains unspeakable in the manifest content of an advertisement. Insofar as

M. Rivera (✉)
Tufts University, Medford, MA, USA
e-mail: miguel.rivera@tufts.edu

C. Burnham, P. Kingsbury (eds.), *Lacan and the Environment*, The Palgrave Lacan Series, https://doi.org/10.1007/978-3-030-67205-8_11

advertisements attempt to produce images of enjoyment, they can only produce either the rational choice of articulable personal preference or the substitute satisfaction of the symptom. This becomes even more significant when considering how the market treats images of the natural world to sell products that inflict tremendous harm on that same space. Even as advertising appropriates psychoanalysis, it can never appropriate *jouissance*.

Advertisers know that a commercial can never depict what it is that would make a consumer buy a product, so they have developed a language to position their product in line with the *objet a* and to suggest that the product is a site of *jouissance*. Anatomizing advertising strategies, techniques, and signs is an important task for understanding the landscape of late capitalism and potentially subverting the powerful impact they can have on human subjects. But understanding these advertising techniques, particularly with a psychoanalytic insight, is also important precisely because of the many ways in which corporations are environmentally exploitative and irresponsible. Climate change scholars such as Mike Hannis have gone so far as to suggest that advertising and marketing are one of the most acute threats to the health of the environment.[3]

Likewise, Jerry Mander considers advertising to be a significant threat to human freedom and flourishing. He writes, "if you accept the existence of advertising, you accept a system designed to persuade and to dominate minds by interfering in people's thinking patterns."[4] Mander points to advertising as the primary mechanism by which human beings and "the environment [are] redesigned into a form that fits the needs of the commercial format."[5] Mander goes on, "advertising is the instrument of transmutation. *It* lays the standard-gauge railway track from wilderness to human feeling, assisting in the transformation of both into a unified commercial format."[6] This is accomplished by separation, both of human beings from each other and of human beings from necessities, such as food and shelter, through a system of commodity capitalism.[7] Pairing Mander's insight with Hannis's claim, we can see precisely how the consumption of single-use commodities, encouraged by advertisement, is disastrously harmful for the environment. Mander is most explicit about this when he writes:

> Advertising has the task of *creating* discontent with what has just been bought.... The newly purchased commodity must be gotten rid of and replaced by the "need" for a new commodity as soon as possible. The ideal world for advertisers would be one in which whatever is bought is used only once and then tossed aside. Many new products have been designed to fit such a world.[8]

In light of Mander's claim, I argue that the limit of what an advertisement can show and how it can appeal to consumers is best understood in the context of Jacques Lacan's formulation of perversion. By considering the perverse subject's relation to fantasy, we can better understand the way in which advertising harnesses desire. DirecTV and SUV advertisements, along with the television series' fictional account of the 1960s advertising firm Sterling Cooper, *Mad Men* (2007–2015), are paradigmatic of the different ways in which advertisements display their inadvertent awareness of the perverse structure and the proliferation of perverse tendencies in ostensibly *normal* subjects and thus are able to arouse the subject's desire.

The 2017 advertising campaign of DirecTV has captured one of the tensions at the core of advertising logic. In the advertisement's text, it boasts "DirecTV has been rated number one in customer satisfaction over cable for 17 years running. But some people still like cable." By metonymy, it subsequently associates the "liking" of cable, demonstrated by the consumer purchase of cable over the alternative of DirecTV's satellite service, with a number of bizarre behaviours. They include banging one's head on a low ceiling, drinking spoiled milk, sleeping in a bush of poison ivy, opening a "preshaken" soda, having one's seat kicked on an airplane, being rammed by a shopping cart, and the most unexpectedly gruesome: getting a paper cut, where the individual cuts their tongue on an envelope intentionally and then laughs gleefully. The commercial closes with, "but for everyone else, there's DirecTV." Everyone else, indeed. Those who are not subject to the whims of desire and thrill of *jouissance*, those completely rational subjects, should surely purchase DirecTV.

And yet, the unexpected gruesomeness (after all, is not the paper cut located on the tongue excessive in a way the ad copy does not suggest in their formulation of "getting a paper cut"?) and the uncanniness of the various banal horrors these commercials depict serve to undercut the notion of the rational consumer. It is the repulsiveness of the acts metonymically associated with purchasing cable that, the advertiser hopes, would drive one to purchase DirecTV over cable rather than the number one customer satisfaction rating that the service touts at the outset. Furthermore, it is a desire that is in and of itself irrational that might compel the subject to disassociate from these violent depictions of enjoying one's symptom. "That's not me," the consumer pleas as they make the supposedly rational choice based only on the need to substantiate a fantasy of rationality. The DirecTV advertisements attempt to reduce "the unruly body of jouissance to the more governable body of pleasure."[9]

It is worth noting that, while these advertisements maintain the illusion of rational choice as the primary impetus for consumer purchase, the apparatus of advertising as a whole seems content with embracing the unconscious wholesale and abandoning even the illusion of courting the rational consumer. In *The Lacanian Left*, Yannis Stavrakakis writes that advertisers, "from the 1950s onwards, having gradually realised that it is not rational argumentation but emotional ties that bind,"[10] began using motivational research techniques developed by Ernest Dichter. Dichter avowed a Freudian influence and built into his motivational research techniques the discovery of the unconscious. Stavrakakis goes on to suggest that this appropriation of psychoanalysis by advertisers accounts for "the analogies between psychoanalytic free association, depth interviews and focus groups."[11] And yet, despite this, there is no question that advertisements, like the DirecTV ads, conflate enjoyment and rationally articulable preference. They efface the gap between "liking" something and the pull of *jouissance*.

Though it is clear by reading the work of advertisers everywhere that television commercials and ad campaigns appeal to the unconscious more than to the rational parts of the ego; there is also an informative fictional representation of 1960s advertising in the wake of psychoanalysis. Consider the scene in the first episode of *Mad Men*, "Smoke Gets In Your Eyes," in which Dr. Greta Guttman suggests that Don Draper produce advertising copy for Lucky Strike involving the "death instinct," following the much-maligned translation of the German *trieb* into "instinct" rather than "drive."[12] Draper says "I find your whole approach perverse," if only he knew, and dismisses the idea out of hand. And yet, is it not an irrational irrelevancy that he embraces to make Lucky Strike appear to stand out? Distinguishing the brand precisely by highlighting something that is uniform across any cigarette. "It's toasted." And what is Draper's closing soliloquy to the Lucky Strike executives? He says, "advertising is based on one thing, happiness. And you know what happiness is? Happiness is the smell of a new car. It's freedom from fear. It's a billboard on the side of the road that screams, with reassurance, that whatever you're doing, it's okay. You are okay." That car driving passed the billboard to which Don refers sounds as if it may be on its own death drive.

But beyond that, what Draper describes is desire's persistence in its attempt to fill the subject's constitutive lack. To say "it's toasted" is not to produce a new desire, but the instillation of a billboard that directs desire towards an object that would offer partial satisfaction as well as any other. If all cigarettes are the same, both as commodities and as sites of temporary, partial satisfaction, one might as well buy Lucky Strikes. It is the fact that desire is persistent, then, that is the engine of advertising. Advertising does not produce desire, but mobilises it.

The desire mobilised by the DirecTV commercials and the Lucky Strike slogan is solicited by different mechanisms. On the one hand, a conceit of rationalism turns the idea of reasoned choice into a symptomatic compulsion. On the other hand, the highlighting of an irrelevancy that is uniform amongst all competing products that in turn gives consumers the alibi they need for libidinal investment in the product. There is a third example that highlights the contradiction of advertising logic and the mobilisation of desire through yet another mechanism: the SUV commercial. The SUV advertisement as a form operates along two registers. It endorses the perverse disavowal and channels the *jouissance* of destruction.

Thinking of the form in general rather than a precise exemplar or particular brand, consider what many of these commercials have in common: a monstrous SUV rolls through an idyllic natural landscape, whether mountain, tundra, or rainforest. Shane Gunster puts it thus in "'You Belong Outside': Advertising, Nature, and the SUV":

> the SUV is routinely depicted in the most spectacular and remote natural locations. These fanciful themes now attract the scorn of many who draw upon them to underscore the rather glaring contradictions between how these vehicles are marketed and how they are actually used: the irony of using pristine images of a hyper-pure nature to motivate the use of a product that consumes excessive amounts of natural resources and emits high levels of pollutants.[13]

The stark disparity between the reality of the SUV's large carbon footprint and the beautiful environs it traverses necessitates the exploration of this ostensible contradiction and its possible resolution beginning with the consideration of the psychoanalytic structure of perversion. Take, for instance, Paul Hoggett's "Climate Change in a Perverse Culture." The SUV advertisement endorses the same perverse disavowal that Hoggett suggests plagues society as a barrier to concerted and sustained action on climate change.

Hoggett defines perversion thus, "perversion involves a turning away from what is true and right or, more strongly, an obstinate or stubborn persistence in what is untrue and unreasonable."[14] Though Lacan is absent from Hoggett's formulation, he describes (with a moralistic valence) the mechanism of disavowal that Lacan identifies as fundamental to the perverse structure. Nonetheless, Hoggett's notion of perversion is distinct from Sigmund Freud's and Lacan's. The perverse structure according to Lacan, in its various permutations, appears "to be a no-holds-barred pursuit of satisfaction by the pervert himself is, in fact … the attempt to bring into being a law that restrains the pervert's jouissance."[15] Stephanie Swales gives an account of the paradigmatic case of perversion, Little Hans, in her text *Perversion: A Lacanian Psychoanalytic Approach to the Subject*, "The fetishist is the child who, unable to assign a signifier to the mother's genitalia, *disavows* the perception of what seems to him the *absence* of a penis and maintains side by side two pieces of contradicting knowledge."[16] This "both/and logic"[17] of perversion is a necessary part of the fundamental irrationality of the human subject that advertisers exploit. Advertisers both indulge and perpetuate this perverse structure in order to sell products using language and imagery that ranges from contradictory to nonsensical. Hoggett makes a harsh moral condemnation of perversion and believes it is something that must be curtailed to combat climate change, whereas Lacan's impression of perversion is as something fundamental to the essence of humanity.[18] Despite the fact that perversion is essential to the sexual practice of humans, relatively few subjects manifest the perverse structure.[19] Hoggett believes that what is at stake is understanding perversion in order to ameliorate it amongst those who reject the validity of climate science. It is not that perversion must be eliminated, as Hoggett suggests, nor that climate deniers possess a perverse structure, but rather that the perverse structure's relationship to fantasy must be understood in relation to advertisements. Advertisers exploit the perversity endemic to the human subject by stimulating desire through fantasy, as is evident in Gunster's analysis of the SUV advertisement.

But Gunster advances his point beyond explicating the contradiction accounted for in psychoanalysis as disavowal. He also suggests that what is at stake in the utilisation of nature is to "make the case that a certain commodity or brand will enable an escape from the malaise and drudgery of urban existence,"[20] in what seems like a distortion of Emersonian self-reliance.[21] This supposedly American characteristic is what this advertisement form attempts to evoke. The SUV is manifest destiny, the obvious phallic implications displaced momentarily from the sports car to the larger, less aerodynamic vehicle. The SUV advertisement represents nature itself as a cite of *jouissance*; but that enjoyment is only accessible inside the car uniquely able to bring one into contact with nature. The *jouissance* that is suggested is partially the product of interrupted routine[22] and freedom through dominance.

One only need to think of the name of Ford's flagship SUV, the Explorer, to understand this point.[23] Gunster takes his analysis yet further when he suggests that the SUV's violence to nature and its ostensible ability to subdue it become part of its appeal.[24] Obliquely, then, it is precisely the ability to destroy nature through the SUV's consumption of natural resources and comparatively large carbon footprint that makes the SUV appealing as a product. Rather than its representation in nature as a contradiction that undoes the advertisement, the tension facilitates the sale of the product. Nature, "cast ... as enemy,"[25] is enjoyed through domination.[26] Each of these means by which the SUV might be a circuit for *jouissance* are located squarely at the level of fantasy, in accordance with the unending *jouissance* that is *only at the level of fantasy* and *rarely actually experienced* by the structurally perverse.[27] Those with a perverse structure experience pleasure derived from their fantasmatic *jouissance* and the possibility of a limit being placed on the possibility of *jouissance* by the instantiation of the law and propping up of the *Nom-du-Pére*.

The *Nom-du-Pére*, Name-of-the-Father, or paternal function is what occasions disavowal. The pervert refuses to give up their own *jouissance* at the behest of the paternal function, but unconsciously desires a limit to be placed on that same *jouissance*. *Mad Men* also offers an exemplary instance of the perverse relation to the environment in the episode "The Gold Violin."[28] The Drapers' appreciation for what nature offers for their relaxation and contemplation—in other words, for their *use*—seems incompatible with their maltreatment of the park as Don launches his empty can like a football and Betty dumps the family's trash from the blanket onto the ground. And yet it is the idea of the natural world as available for use and exploitation, the mutually exclusive possibilities of the park remaining in all its *natural splendour* and also being available as a dumping ground, that make this behaviour possible. Don and Betty engage in an obvious act of perverse disavowal.[29] When Betty suggests "we should do this more often" and Don replies "we should only do this," what the pair *actually* describes is the destruction of nature rather than the retreat into it for an afternoon of relaxation. Don, in fact, "only" works in service of the environment's annihilation in his role as an advertiser.

The question of the pervert's relation to fantasy is at odds with *jouissance* as a product of the drive, though both the SUV advertisement and the DirecTV advertisement deal with each to varying degrees. Even without engaging with Lacan explicitly, in "Unconscious Obstacles to Caring for the Planet," John Keene's essay captures the careful negotiation of *jouissance* reflected in the SUV advertisement when he writes, "Pleasure in destruction provides quicker and more immediate satisfaction than the painstaking matter of creation or repair of things. This is sadly a frequent gratification for politicians and managers where the glee in inflicting pain can be only too obvious. It is particularly worrying where the environment is the unconscious target."[30] Keene's observation elaborates Gunster's point of nature cast as enemy. Keene also offers a harsh moral condemnation of what one might identify as the drive and, perhaps, an all too quick association with the "pleasure" one might take in "destruction," depletion, or consumption with the sadism of certain political figures. Furthermore, the word "pleasure" is, in my view, an incorrect description of what Keene identifies—what Keene is identifying is in fact *jouissance*.[31] Although Lacan suggests in *Seminar VII*, "we cannot avoid the formula that *jouissance* is evil,"[32] it is precisely the point that, in advertisements that evoke *jouissance*, enjoyment and morality are disentangled or their relation is recalibrated. Often that recalibration means enjoyment is positioned as the highest good and aspiring towards it the only worthwhile task, whatever the cost.

Just as Lacan makes a claim about the ethical status of *jouissance* in *Seminar VII*, to be expected in a seminar subtitled *The Ethics of Psychoanalysis*, Lacan also offers a moral imperative that he believes to be at the core of psychoanalysis at the seminar's conclusion: "the only thing of which one can be guilty is of having given ground relative to one's desire."[33] The ethical principle, then, that governs psychoanalysis—the claim that one must never give ground relative to one's desire—is the principle that corporate interests attempt to exploit for their own gain by *redirecting* desire towards a product through an advertisement. Lacan goes on:

> If analysis has a meaning, desire is nothing other than that which supports an unconscious theme, the very articulation of that which roots us in a particular destiny, and that destiny demands insistently that the debt be paid, and desire keeps coming back, keeps returning, and situates us once again in a given track, the track of something that is specifically our business.[34]

The insistent, incessant return of desire is the most powerful compulsive force, so no wonder corporations want to exert some level of control over it. By moulding, changing, redirecting one's desire, they may change that individual's very destiny.

This form of enjoyment, the indulgence of one's desire at any cost desublimated into the purchasing of a consumer product, also brings to mind the discourse of the DirecTV advertisement and the sublimation of drive into rationality through metonymic representation of symptoms. No matter how often the language of the advertisement describes the qualities of a product, those qualities are *never* what the corporation actually believes will make someone purchase the product. What Keene identifies as pleasure and I identify as *jouissance* is absolutely at odds with rationality. John Steiner, in a response to Hoggett's essay, connects the issues of perversion and pleasure to rationality and, in his words, the "death instinct."[35] He writes, "Even though we are aware of unconscious mechanisms and destructive forces operating in the setting of the consulting room, many psychoanalysts, and I include myself, regularly expect people to behave rationally and decently in the world at large, and can become outraged when they do not."[36] Perhaps one of the most agreed upon insights among all psychoanalytic orientations is the fantasmatic status of the rational subject. Keene endorses Steiner's claim, writing, "our public discourse tends to follow the Enlightenment view that rational thought now predominates, and there is recurrent surprise on finding so frequently that this is not the case."[37]

While I strongly dispute the ethical evaluation of the death drive[38] posited by both Keene and Steiner, their identification of the drive as inextricably linked to issues of climate change and perversion is thought provoking. I would argue that their insights have more relevance to the function of the advertisement writ large as a means by which climate change is aided and abetted through restructuring human behaviour.

Like the DirecTV advertisement, the SUV advertisement stops short of being able to depict *jouissance* in a way that is legible to the viewer. Instead, entering the natural space via the SUV is represented as a rational choice, undergirded by the litany of features outlined in the advertisement. The SUV ad also covers over the gap between avowed preference and disavowed desire. Furthermore, this tradition of advertising affirms nature as a fantasmatic signifier that can cover over the subject's constitutive lack somehow exacerbated by urban living. Gunster makes much of this signifier's existence as the "problem" of the SUV advertisement, rather than the existence of the SUV itself. Gunster seems to be advancing the thesis of someone like Mei Mei Evans[39] or Timothy Morton[40] who both suggest that the signifier of nature is socially contingent. Lacan, himself, seems to prefigure this thesis as well. When speaking at Sainte-Anne Hospital in 1971, he says, "Let's speak about the natural. It is everything that is clad in the livery of knowledge, and Lord knows there is no dearth of it.... Don't imagine that I'm one of those who pit culture against nature, if only because nature is precisely a fruit of culture."[41]

Just as Lacan anticipates the SUV commercial's deployment of nature as signifier, he makes obvious why the depiction of *jouissance* as such is difficult or impossible. Bruce Fink writes in *Lacan to the Letter*, "Why is it the Other jouissance cannot be spoken? If it were spoken, it would have to be articulated in signifiers, and if it were articulated in signifiers, it would be subject to the bar between signifier and signified."[42] Though Fink here refers specifically to the "Other *jouissance*" or feminine *jouissance* as opposed to phallic *jouissance* or masculine *jouissance*, Lacan leaves absolutely no ambiguity about this point in "Subversion of the Subject and Dialectic of Desire." Lacan writes, "we must insist that *jouissance* is forbidden to him who speaks as such, although it can only be said between the lines for whoever is subject of the law, since the law is grounded in this very prohibition."[43] *Jouissance* is prohibited by the law and given up by the subject to enter the symbolic universe. My claim is not simply that *jouissance* eludes representation in advertising media but precisely that it is the conflation of the absented *jouissance* and the represented reasoned choice that advertisers deliberately exploit. The tension between the two, the negative space of *jouissance* in discourse and depictions of reasoned choice, is how an advertisement, functioning in discourse, is able to "get its hooks in" so to speak. That tension itself, just like the contradiction between the SUV and the signifier of the natural world, provides a certain type of enjoyment. Lacan writes of tension, "Yet, why enjoy, if not because it produces a tension? This tension is the very principle of everything that goes by the name of jouissance."[44] Stavrakakis also reinforces this point when he articulates the "enjoyment entailed in consuming advertisements."[45] The momentary substitute *jouissance* of the advertisement initiates the metonymic approach of the consumer towards the commodity.

What these advertisements all have in common is the revelation of the limits of rationality. These advertisements suggest the possibility of rational choice, weighing the pros and cons, checking the data, making the wise decision, and yet ultimately foreclose that possibility entirely. After all, where is the space for rational choice in a world where people "like," or enjoy drinking spoiled milk or sitting in gum? Can the advertisement honestly suggest that the viewers are so exceptional, they share none of these bizarre quirks in themselves? Is it not a quirk that these advertisers hope for that might draw someone to a satellite dish or SUV? The DirecTV advertisements in particular align the continued purchasing of cable with the pathological behaviours that they stop short of naming as such. However, what is erroneous in the construction of their advertising fantasy is that there is a rational alternative, someone immune to the seduction of *jouissance* that might cause one to enjoy drinking sour milk.

These DirecTV commercials prohibit particular kinds of enjoyment and prioritize over them the preference of reasoned choice. However, it is in turn the fantasy structure of reasoned choice that sustains libidinal investment, and thus in itself becomes irrational. This turn is similar to that of the SUV commercial. The subject perversely disavows nature's destruction as the SUV cruises through the forest or across mountain tops. The advertisement's perverse structure and its dismantling of capacities of rational choice pose obvious problems for addressing issues of climate change, especially when climate change activism often makes its arguments using statistics and research that only a completely rational subject would respond to. This kind of argumentation is powerless against the perverse belief that the only way to *really* enjoy nature is from the front seat of a Jeep.

These advertisements also participate in a recalibration of ethics in relation to consumption and appropriate Lacan's dictum from *The Ethics of Psychoanalysis*, one must never give ground relative to one's desire. Because desire is difficult to represent in signifiers, and *jouissance* is entirely unrepresentable, what is depicted is often the striking negative space of *jouissance*'s absence or perhaps a substitute *jouissance*, sometimes derived from the superego. Even the ethical gesture—the condemnation of other products' corporate policies, environmental impact, or gender politics—can offer some facsimile of *jouissance*. This is evident especially in television advertising today. As this new type of advertising strategy emerges that positions a company as moral and thus suggests their products should be purchased because of the policies of the company, perhaps the polar opposite of the SUV advertisement that depicts the vehicle in nature, we can return to *Mad Men* for another informative example. In the twelfth episode of the fourth season, "Blowing Smoke," one can find a version of a righteous, superegoic substitute *jouissance* in the instance of Don Draper condemning the cigarette companies for which he used to write copy. And, of course, Draper's repudiation of the tobacco industry is an advertisement in and of itself.[46]

As Draper demonstrates, the apparatus of advertising wants to know everything about *jouissance*. Of course, whatever knowledge of *jouissance* one might want or not want, the relation between knowledge and *jouissance* remains antagonistic. *Jouissance* cannot be known. But still, advertising is always grasping at the drive. Draper understands this and points the finger at his colleagues who do not as they critique his "Why I Quit Tobacco" advertisement, "It's an ad for this agency. If you don't understand it, then you shouldn't be in this business." But *jouissance* is always absent from the picture, the symbolic, the discursive frame that advertising generates. Don again intuits this when he encounters his old flame Midge and comments on her heroin addiction. Don, deceived by Midge and brought to her apartment in an effort to sell him one of her paintings, is able to see her neglected living space and becomes aware of her financial destitution as a result of the addiction. Don asks Midge questions such as "what's it like?" and "why don't you just quit?" After Midge gives an account of her experience under the influence of heroin, Don looks around the squalid apartment and says "I can see it's very good." What it is Don sees, then, is precisely the *nothing* of *jouissance*, the absent presence it precipitates and necessitates. The *jouissance* of a heroin high can never be conveyed to Don in a language or form he would understand, not Midge's description of the experience of the drug and certainly not her "Number 4" painting. And yet, what gives Don insight into that *jouissance* is the absence of everything else: possessions, money, meaningful relationships, and self-control.

As Don turns his back on Lucky Strike and the cigarette companies that, in the words of Burt Cooper, "put a roof over [his] head and … fed [his] children," it should be clear the way in which *Mad Men*'s engagement with smoking offers powerful insights into the way advertising functions and also works as an effective allegory for the question of climate change and corporate responsibility. After all, are carbon emissions not simply a broader instantiation of the act of individual cigarette consumption? In both cases, armies of lobbyists are mobilised to cover over any negative consequences of the act and legions of advertising agencies work to divest people of their sense and channel their desire into a product that may be harmful, either to the Earth or to one's own body. If nothing else, the consistent structure of the workings of desire and *jouissance* in relation to these two issues makes them invaluable foils. The same can be said examining the discourses of the advertisements for DirecTV and the SUV. Although they are distinct, the question of how advertisements function must be answered to understand the individual and societal relation to the issue of climate change. Because of the psychic structure of perversion and the power of desire and *jouissance*, it is easy for advertisers to murky the issue of the connection between one's own consumption and climate change. An awareness of the advertising strategies that appropriate these psychoanalytic principles helps us live up to Lacan's dictum, to never give ground relative to our desire, instead of putting our desire in service of environmentally harmful corporate interests.

Notes

1. Yannis Stavrakakis writes, "Fantasy is a construction that stimulates or causes desire because it promises to cover for the lack created by the loss of *jouissance* with a substitute, a miraculous object, the *objet petit a*" in *The Lacanian Left: Psychoanalysis and Politics* (Edinburgh: Edinburgh University Press, 2007), 239.
2. "If advertising attempts to stimulate, to cause our desire, this can only mean that the whole mythological construction it articulates around the product is a social fantasy and, furthermore, that this product serves or functions as an object that causes desire, in other words as an object-cause of desire, an *objet petit a* in the Lacanian vocabulary" (Stavrakakis, *Lacanian Left*, 238). See also Jean Baudrillard, *The Consumer Society: Myths and Structures* (London: Sage, 1998), 33.
3. Mike Hannis writes, "Restrictions on marketing, broadly construed, would surely be among the most effective climate change mitigation

measures available. They would, of course, also be good for our own mental and physical health" Hannis, "Discussion: Nature, Consumption and Human Flourishing," in *Engaging with Climate Change: Psychoanalytic and interdisciplinary perspective*, Sally Weintrobe ed. (London: Routledge, 2013), 218.

4. Jerry Mander, *Four Arguments for the Elimination of Television* (New York: Morrow Quill, 1978), 45.
5. Mander, *Four Arguments*, 120.
6. Mander, *Four Arguments*, 125; emphasis in the original.
7. Mander writes, "advertising intervenes between people and their needs, *separates* them from direct fulfillment and urges them to believe that satisfaction can only be obtained through commodities" (*Four Arguments*, 127; emphasis in the original).
8. Mander, *Four Arguments*, 129; emphasis in the original.
9. Colin Wright, "Perversion and the Law: From Sade to the 'Spanner Case' and Beyond," in *Perversion Now!*, ed. Diana Caine and Colin Wright (Cham, Switzerland: Palgrave Macmillan, 2017), 229.
10. Stavrakakis, *Lacanian Left*, 232.
11. Stavrakakis, *Lacanian Left*, 232.
12. Matthew Weiner, "Smoke Gets in Your Eyes," *Mad Men*, created by Matthew Weiner, directed by Alan Taylor, Season 1, July 19, 2007 (Los Angeles: American Movie Classics, 2007). Jacques Lacan writes, "The drive—the Freudian drive—has nothing to do with instinct" "On Freud's 'Trieb' and the Psychoanalyst's Desire," in *Écrits: The First Complete Edition in English*, trans. Bruce Fink (New York: W. W. Norton, 2006), 722.
13. Shane Gunster, "'You Belong Outside': Advertising, Nature, and the SUV," *Ethics and the Environment* 9, no. 2 (2004): 4, https://www.jstor.org/stable/40339086.
14. Paul Hoggett, "Climate Change in a Perverse Culture," in *Engaging with Climate Change: Psychoanalytic and Interdisciplinary Perspectives*, ed. Sally Weintrobe (London: Routledge 2013), 59.
15. Bruce Fink, *A Clinical Introduction to Lacanian Psychoanalysis: Theory and Technique* (Cambridge, MA: Harvard University Press, 1997), 192.
16. Stephanie S. Swales, *Perversion: A Lacanian Psychoanalytic Approach to the Subject* (London: Routledge, 2012), 69.
17. Swales, *Perversion*, xiii.

18. Lacan posits, "all human sexuality is perverse if we carefully follow what Freud says" in *The Seminar of Jacques Lacan, Book XXIII: The Sinthome* (2005), ed. Jacques-Alain Miller, trans. A. R. Price (Cambridge: Polity Press, 2016), 132.
19. Fink, *Clinical Introduction*, 165.
20. Gunster, "You Belong Outside," 4.
21. Gunster writes, "Ads wax poetically about the quiet virtues of isolation in contrast to the crowded, noisy streets of the city. 'I never found the companion so companionable as solitude,' notes a Chevrolet Blazer ad, approvingly quoting the words of Henry David Thoreau" ("You Belong Outside," 7).
22. This is, perhaps, also the *jouissance* of colonialism. Colonialism serves as a structural analogy to the SUV advertisement: to enjoy something is also to annihilate it.
23. Gunster, "You Belong Outside," 6.
24. Gunster, "You Belong Outside," 9.
25. Gunster, "You Belong Outside," 9.
26. Gunster goes on to claim, quite rightly, that it is the signifier of nature itself that is the problem rather than the advertising strategies employed. That is to say, the image of nature as a pristine escape from the harsh urban landscape is a fantasy that enables advertising. To say that the SUV must be criticised or refused because nature must be protected is only to affirm the consistency of this fantasmatic signifier ("You Belong Outside," 9).
27. Fink, *Clinical Introduction*, 165.
28. Written by Jane Anderson, Andre Jacquemetton, and Maria Jacquemetton, "The Gold Violin," *Mad Men*, created by Matthew Weiner, directed by Andrew Bernstein, Season 2, September 7, 2008 (Los Angeles: American Movie Classics, 2008).
29. Likewise, when Don tells his son Bobby to urinate behind a tree and his daughter Sally subsequently says, "I want to tinkle outside," one can immediately think of the specific circumstances of the Little Hans case with Sally blissfully unaware of what is at stake in her question (*Perversion* 69).
30. John Keene, "Unconscious Obstacles to Caring for the Planet: Facing Up to Human Nature," in *Engaging with Climate Change: Psychoanalytic and Interdisciplinary Perspective*, ed. Sally Weintrobe (London: Routledge, 2013), 147.

31. For Lacan, pleasure and *jouissance* are mutually exclusive and antagonistic towards one another because *jouissance* entails traversing the pleasure principle. Lacan writes, "For it is pleasure that sets the limits on *jouissance*" "Subversion of the subject and dialectic of desire in the Freudian unconscious" in *Écrits: A Selection*, trans. Alan Sheridan (New York: W. W. Norton, 1977), 319.
32. Jacques Lacan, *The Seminar of Jacques Lacan, Book VII: The Ethics of Psychoanalysis, 1959–1960*, ed. Jacques-Alain Miller, trans. Dennis Porter (New York: W. W. Norton, 1992), 184.
33. Lacan, *Écrits*, 319.
34. Lacan, *Écrits*, 319.
35. Steiner's use of the phrase "death instinct" demonstrates the difference between his "post-Kleinian" and the Lacanian orientation. As I have noted, the Lacanian position would be to maintain a distinction between Freud's use of *trieb* (perhaps more accurately translated as "drive," but translated by James Strachey as "instinct") and *instinkt* ("instinct," referring to biological urges). For the purposes of my work, I will refer to what Steiner calls the "death instinct" as the death drive.
36. John Steiner, "Discussion: Climate Change in a Perverse Culture," in *Engaging with Climate Change: Psychoanalytic and interdisciplinary perspective*, ed. Sally Weintrobe (London: Routledge, 2013), 81.
37. Keene, "Unconscious Obstacles," 144.
38. "The life instinct creates structure and meaning, and the death instinct attempts to destroy meaning and to create an amorphous uniformity in which no structure is discernible. It is really more accurate to consider the death instinct as an anti-life instinct, which is most vividly expressed through destructive attacks on goodness, truth and creativity" (Steiner, "Discussion," 82).
39. Mei Mei Evans writes, "Not only is the hegemonic concept of Nature a masculinist social construction, but one that is racist and heterosexist as well" "'Nature' and Environmental Justice," in *The Environmental Justice Reader: Politics, Poetics, and Pedagogy*, ed. Joni Adamson, Mei Mei Evans, and Rachel Stein (Tucson: University of Arizona Press, 2002), 191.
40. Timothy Morton writes, "We are losing touch with a fantasy Nature that never really existed" "Guest Column: Queer Ecology," *PMLA* 125, no. 2 (2010): 273, https://doi.org/10.1632/pmla.2010.125.2.273.

41. Jacques Lacan, *Talking to Brick Walls: A Series of Presentations in the Chapel at Sainte-Anne Hospital*, trans. A. R. Price (Cambridge: Polity Press, 2017), 30–31.
42. Bruce Fink, *Lacan to the Letter: Reading Écrits Closely* (Minneapolis: University of Minnesota Press 2004), 162.
43. Lacan, "Subversion of the Subject and Dialectic of Desire," in *Écrits*, 319.
44. Lacan, *Talking to Brick Walls*, 23.
45. Stavrakakis, *Lacanian Left*, 243.
46. Andre Jacquemetton and Maria Jacquemetton, "Blowing Smoke," *Mad Men*, created by Matthew Weiner. directed by John Slattery, Season 4, October 10, 2010 (Los Angeles: American Movie Classics, 2010).

Bibliography

Anderson, Jane, Andre Jacquemetton, and Maria Jacquemetton. 2008. The Gold Violin. *Mad Men*. Created by Matthew Weiner. Directed by Andrew Bernstein. Season 2. September 7, 2008. Los Angeles: American Movie Classics.

Baudrillard, Jean. 1998. *The Consumer Society: Myths and Structures* (1970). London: Sage.

DIRECTV TV Commercial. 2017a. Head Bang. *iSpot.tv*. Ad ID: 1621166. Accessed September 22, 2019. http://www.ispot.tv/ad/wlDi/directv-head-bang.

———. 2017b. Pre-Shaken Soda. *iSpot.tv*. Ad ID: 1659751. Accessed September 22, 2019. http://www.ispot.tv/ad/wEZe/directv-pre-shaken-soda-reward-card.

———. 2017c. Wet Bags. *iSpot.tv*. Ad ID: 1632928. Accessed September 22, 2019. http://www.ispot.tv/ad/wrKR/directv-soggy-groceries.

Evans, Mei Mei. 2002. 'Nature' and Environmental Justice. In *The Environmental Justice Reader: Politics, Poetics, and Pedagogy*, ed. Joni Adamson, Mei Mei Evans, and Rachel Stein, 181–191. Tucson: University of Arizona Press.

Fink, Bruce. 1997. *A Clinical Introduction to Lacanian Psychoanalysis: Theory and Technique*. Cambridge, MA: Harvard University Press.

———. 2004. *Lacan to the Letter: Reading Écrits Closely*. Minneapolis: University of Minnesota Press.

Gunster, Shane. 2004. 'You Belong Outside': Advertising, Nature, and the SUV. *Ethics and the Environment* 9 (2): 4–32. https://www.jstor.org/stable/40339086.

Hannis, Mike. 2013. Discussion: Nature, Consumption and Human Flourishing. In *Engaging with Climate Change: Psychoanalytic and Interdisciplinary Perspectives*, ed. Sally Weintrobe, 214–220. London: Routledge.

Hoggett, Paul. 2013. Climate Change in a Perverse Culture. In *Engaging with Climate Change: Psychoanalytic and Interdisciplinary Perspectives*, ed. Sally Weintrobe, 56–71. London: Routledge.

Jacquemetton, Andre, and Maria Jacquemetton. 2010 Blowing Smoke. *Mad Men*. Created by Matthew Weiner. Directed by John Slattery. Season 4. October 10, 2010. Los Angeles: American Movie Classics.

Keene, John. 2013. Unconscious Obstacles to Caring for the Planet: Facing Up to Human Nature. In *Engaging with Climate Change: Psychoanalytic and Interdisciplinary Perspectives*, ed. Sally Weintrobe, 144–159. London: Routledge.

Lacan, Jacques. 1977. The Subversion of the Subject and the Dialectic of Desire in the Freudian Unconscious (1960). In *Écrits: A Selection*. Translated by Alan Sheridan, 292–325. New York: W. W. Norton.

———. 1992. *The Seminar of Jacques Lacan, Book VII: The Ethics of Psychoanalysis, 1959–1960* (1986). Edited by Jacques-Alain Miller. Translated by Dennis Porter. New York: W. W. Norton.

———. 2006. On Freud's 'Trieb' and the Psychoanalyst's Desire (1964). In *Écrits: The First Complete Edition in English*. Translated by Bruce Fink, 722–25. New York: Norton and Norton, 1996; Reprint. W. W. Norton.

———. 2016. *The Seminar of Jacques Lacan, Book XXIII: The Sinthome* (2005). Edited by Jacques-Alain Miller. Translated by A. R. Price. Cambridge: Polity Press.

———. 2017. *Talking to Brick Walls: A Series of Presentations in the Chapel at Sainte-Anne Hospital* (2011). Translated by A. R. Price. Cambridge: Polity Press.

Mander, Jerry. 1978. *Four Arguments for the Elimination of Television*. New York: Morrow Quill.

Morton, Timothy. 2010. Guest Column: Queer Ecology. *PMLA* 125 (2): 273–282. https://doi.org/10.1632/pmla.2010.125.2.273.

Stavrakakis, Yannis. 2007. *The Lacanian Left: Psychoanalysis and Politics*. Edinburgh: Edinburgh University Press.

Steiner, John. 2013. Discussion: Climate Change in a Perverse Culture. In *Engaging with Climate Change: Psychoanalytic and Interdisciplinary Perspectives*, ed. Sally Weintrobe, 80–83. London: Routledge.

Swales, Stephanie S. 2012. *Perversion: A Lacanian Psychoanalytic Approach to the Subject*. London: Routledge.

Weiner, Matthew. 2007. Smoke Gets in Your Eyes. *Mad Men*. Created by Matthew Weiner. Directed by Alan Taylor. Season 1. July 19, 2007. Los Angeles: American Movie Classics.

Wright, Colin. 2017. Perversion and the Law: From Sade to the 'Spanner Case' and Beyond. In *Perversion Now!* ed. Diana Caine and Colin Wright, 221–230. Cham, Switzerland: Palgrave Macmillan.

Žižek, Slavoj. 2012. *Less Than Nothing: Hegel and the Shadow of Dialectical Materialism*. London: Verso.

Part IV

End of the World

12

Psychoanalysis at the End of the World

Nathan Gorelick

Here on the verge of oblivion, we peer like timid children over the edge of the great abyss our parents industriously carved from our future. As ever, as on smaller scales, we love them and we hate them for it. We want to be loved by them, to be worthy of this terrible inheritance, to have earned the world we have been given, with all its decrepitude and the mounting treasury of its disaster; and we want our freedom from this obligation to be loved, the freedom to refuse this awful bequest on the way to another world, different and of our own making, since surely it is better to be responsible for one's own problems than to assume somebody else's. Standing at this edge, eyes wide with wonder and fear and, if we are being honest, a little excitement, we imagine the worst and hope for the best.

There is no more fitting discourse with which to think this condition and the crisis of responsibility it entails than psychoanalysis. From the beginning, the Freudian discovery was built upon the question of infantile ambivalence and its consequences, as much for every subject as for civilisation and, indeed, our species. This is at least implicitly the starting

N. Gorelick (✉)
Utah Valley University, Orem, UT, USA
e-mail: nathan.gorelick@uvu.edu

C. Burnham, P. Kingsbury (eds.), *Lacan and the Environment*, The Palgrave Lacan Series, https://doi.org/10.1007/978-3-030-67205-8_12

point for existing psychoanalytically informed investigations of environmental degradation and its discontents, such as, for example, Renee Lertzman's *Environmental Melancholia*, which considers the disconnect between people's stated beliefs regarding environmental collapse and their contrary actions perpetuating the very problems about which they purport to care. Through this lens, what reads as public apathy in the face of ecological catastrophe is only the outwardly measurable indication of a paralysing anxiety which, properly understood, explains why "people often avoid what is most painful, which ecological threats arguably are, even when it is counterproductive." The horizon of this intervention is "a cogent environmental communications theory and practice" that wants to overcome anxiety on the way to meaningful action.[1] To bear all this out, however, we first must contend with the blind spot in Lertzman's analysis: the kernel of *jouissance* to which this anxiety attests—and, with it, the defensive function at play in the fabricated illusion of the end-to-come—in order to track the disavowed real, the truth of the illusion that the world is not already over. In other words, we have to consider our ambivalence and the anxiety it generates according to the logic of the symptom around which representations of the end of the world are organised.

From this Lacanian perspective, the representational field of our anxiety does not fail to signify so much as it signifies otherwise, testifying to an impasse in the order of the signifier which is the mark of the unconscious at work. In what follows, I explore this symptomatic dimension of the end of the world by way of Hollywood's apparent obsession with apocalypse, because this exemplifies the tendency to hope for the best while imagining the worst. Confining our attention to this domain of popular culture permits us to see most clearly how this ambivalence is not a theoretical conceit but a global phenomenon, one which has been energised and incentivised by the immense corporate profits it generates and which, moreover, has the structure of a repetition-compulsion.

By thus isolating the matter, we will find that psychoanalysis affords a vital critical and ethical sensibility with which to navigate—perhaps even undertake—the end of the world as an agonising but necessary condition for the possibility of any world to come. Following Sigmund Freud and Jacques Lacan along the trail of the symptom on the way to the fantasy that structures it, we might resolve the impasse in the symbolic order Lertzman correctly diagnoses as an instance of collective melancholia. Because the only way to move beyond melancholic anxiety is by mourning what has been lost, psychoanalysis can help unbind us from the exhausted world and release the possibility of another future—not an illusory future in which we can repair the irreparable, but the real, imminent, and inevitable future in which we will have to gather the ruins of the present and construct another world out of what remains.

A False Sense of Occasion

The repetition-compulsion at work in the endless outpouring of popular representations of the end of the world is nowhere better figured than in Douglas Adams' second instalment of his four-part *Hitchhiker's Guide* trilogy, *The Restaurant at the End of the Universe*. Night after night, the eponymous restaurant's clientele sip martinis and watch from the literal margins of time and matter as the universe implodes. At the conclusion of every dinner service, the restaurant's "time turbines" rock it back to the other side of the big crunch, where it prepares for the next evening's guests. The fortunate patrons pay for the extravagant show with billions of years of compound interest accrued from a penny placed in a bank account during whichever time they usually reside. Reservations are always tough. "The end of the universe is very popular," one character explains; "People like to dress up for it.... Gives it a sense of occasion."[2] If, as both Slavoj Žižek and Fredric Jameson have noted, it is easier to imagine the end of the world than the end of capitalism, Adams helps us see why: because the end of the world is an especially desirable commodity.

This insight is already at stake in Žižek and Jameson's axiom, although, like any great axiom, it is so often repeated and has become so repeatable that the truth it means to convey has been obscured.[3] The point is not merely that capitalism constricts our vision of the future to the parameters of possibility that ensure the reproduction of its own conditions—as was already made clear, for example, in Louis Althusser's critique of ideology.[4] The point is that capitalism already is the end of the world—protracted or accelerated, obvious or subtle, accomplished or in process depending on one's point of view or social location, but no less terminal for all that. We all know this, even or especially if, to borrow the litany with which Lacan inaugurates his *Encore* seminar, we "don't want to know anything about it."[5] This is why the endlessly repeated apocalypse, far from connoting a failure of political imagination, is a symptom. In other words, and as Adams makes clear, the popular spectacle of the end of the world indexes not a desire for such an end but a contrary attachment to the very world we enjoy watching collapse or explode again and again—a dream (that is, a wish-fulfilment) that the world will carry on, safely insulated from catastrophe by the latter's amusing entrapment within the entertainment media that commoditise it. It is at once a defense against and an acknowledgement of capitalism's cannibalistic orientation towards its own conditions of possibility, an inverted expression of the truth that capitalism is the destruction of the world.

Jean Baudrillard observes as much, and in conveniently Lacanian terms, through his diagnosis of the "catastrophic imaginary." Within this frame, capitalism compels us to imagine the end of the world as a singular catastrophic event, always on the horizon, in the storm clouds gathering just at the edge of our quotidian reality, because this situates the catastrophic present and the catastrophised future in two parallel universes. In this way, the catastrophic imaginary defers any confrontation with the dire consequences of the world economic order—and, of course, any responsibility for those consequences—to an ever-receding future with which we will never intersect.[6]

In the case of the end of the world, the catastrophic imaginary is processed and its parallelism maintained through two common formulas, which I will call the *messianic-redemptive* and the *naturalistic-sentimental.* Their commonality and iterability result from their correspondence to the two halves of our modernity, the theological and the positivist/empiricist, which, contrary to what the modern myth of secularisation would have us believe, are not oppositional but complementary. In Lacanian terms, our symptomatology reveals that both formulas are organised by a shared logic of fantasy, an orientation towards the real, such that their apparent antagonism is merely imaginary. As we will see, the practical horizon of this logic—that is, its result in the (catastrophic) imaginary—is an anthropocentric pretention to mastery, an illusion of human exceptionalism, to which the ethics of psychoanalysis and its injunction to refuse all such pretentions offers the most rigorous rebuke.[7]

To arrive at this conclusion, we should take both types in turn, beginning with the messianic-redemptive.

The eschatological lines in the Hollywood Code are clearly written according to a blueprint of apocalypse as indiscriminate extermination, the wrathful vengeance of a Nature betrayed or at least forgotten. The variables are different but the formula is always the same: *Armageddon, Independence Day, The Fifth Element, The Day After Tomorrow, 2012, Deep Impact* (the saving spaceship in this film is actually named "Messiah"), *Geostorm*, and so on, all follow a basic Biblical pattern—Old Testament judgment, New Testament salvation—modulated by an essentially protestant rationality that substitutes religious devotion and divine election for a few more apparently secular/materialist values. The structure, that is, remains in place; only its signifiers change. Almighty Nature replaces an omnipotent God.[8] The doomsayer-prophet is transmuted into a plucky scientist or two whose special sensitivity permits them to read the signs of Nature's terrible retribution, though not the means to our salvation. That last miracle belongs to the action hero: the grizzled veteran, destitute farmer, or cop who lost it all in some prior act of injustice, but in any case the down-and-out everyman, often an alcoholic, always a reluctant recruit, pulled from his humble origins and thrust into the greatest drama the world has ever known (until the next one comes along), where he redeems himself and his world through an act of suicidal sacrifice that imbues the world's salvation with a messianic significance.

The historical and formal prototype for the nexus between the messianic-redemptive and the naturalistic-sentimental is Daniel Defoe's 1719 novel *Robinson Crusoe*, which is less an adventure tale than a protestant conversion narrative, a didactic yarn which places the hero in the shadow of an invisible God of whom Crusoe's rational ingenuity is a faithful reflection.[9] This is the modern myth and blueprint for the wide variety of post-apocalyptic dramas of which the zombie apocalypse is presently the most popular iteration. Like *Robinson Crusoe*, the zombie motif involves a solitary individual joined by fate to a small, vulnerable group, all cast ashore upon the coast of an uncertain future, collecting the scraps and remnants of the past in order to defend against a hostile or poisoned world, preserving or reviving bourgeois kinship patterns as they search for meaning through the sentimentalism of sometimes romantic, sometimes familial interpersonal relationships. At the level of narrative structure, Crusoe's mindless, God-forsaken cannibals from whom he rescues his Man Friday can be substituted with the rage-infected masses in *28 Days Later*; roving hordes of manic, bloodthirsty goons, as in portrayals of environmental apocalypse like *The Road* or *Mad Max* or *Waterworld*; the bare life who, having left Earth to its total ecological collapse, stuff their oversized faces with junk food on endlessly circulating off-world pleasure cruises in *WALL-E*; soulless, ruthless computer programs like the Agents in *The Matrix*; and on and on.[10] The messianic dimension of the hero's journey is often maintained here while the imagined scale of the world is reduced to more intimate (thus sentimental) proportions; the world as such is replaced with the unlikely family, usually anchored by a heterosexual romance and a helpless, often orphaned, child in whose future the reluctant hero finds his redemptive purpose.

If the catastrophic imaginary is a key preservative of the illusion that the real catastrophe has not happened and never will, then what, precisely, is being preserved, in however inverted a form, by these ever-proliferating variations on the end of the world, and why?

Freud tells us that the theological structure of the messianic-redemptive is a topography whose strata are less layered than superimposed, each new layer retaining and complicating its precedents. The Christian stratum (for Freud, especially its Catholic variant, which despite its divergence from the more Protestant structure we are tracing here is nevertheless the latter's historical precedent and scaffolding) thus is an animistic modulation of an older, transcendental monotheism. Its Trinitarian belief structure is the revival of an ancient polytheism that preceded the Judaic doctrine of an abstract, all-powerful God, a revival that yet retains this same doctrinal abstraction.[11] Through this modulation, the divine becomes less punitive than benevolent, and is imbued with a penchant more for grace than jealous wrath. The mythic structure of the sacrificial son through whose selflessness the world is saved and death finally overcome thus represents a wish-fulfilment, a collective manifestation of the infantile fantasy of the protective, benevolent, all-powerful father, joined with a fantasy of mastery, of total control over all that is and could be, over life and death. In short, it is the fantasy of a union between father and son, of a being without lack.[12]

In the mediatic domain of the catastrophic imaginary, this fantasy is displaced from its religious origins and projected upon what Freud elsewhere (in a thinly disguised allusion to *Robinson Crusoe*) calls "His Majesty the Ego," personified in the heroic everyman with whom the audience is encouraged to identify.[13] The transcendental and the animistic maintain their union while the object of their adoration, the benevolent Father, is supplanted by a Nature to which the hero genuflects even as he endeavours to dominate it. The laboratory is superimposed upon the temple, the scientist upon the figure of the high priest. The archaic, infantile desire for absolute mastery now shrouds itself in the language of rationalism and materialism. Nietzsche was right, God is dead and we killed him, yet we nevertheless continue to dwell in the house we built for Him; in God's place now stands a new master signifier, Nature, the mastery of which holds our only hope for salvation or, what amounts to the same thing in this context, mere survival.[14]

The romanticised or sentimental structure of the social bond, meanwhile, with which the formula is so often saturated, preserves the set of values—the very conception of value as such—in which this Protestantised messianic fantasy historically unfolded and because of which it is so frequently repeated. Individualism and the imperatives of self-preservation, modified by the basic unit of bourgeois moral and economic order (the family), operating within an essentially patriarchal frame, thus continue to monopolise the parameters of our collective imagination as far as concerns species survival, if not happiness. This is why, as we all know—and if you were as yet unaware, please take note—the first and best place to seek refuge in case of a zombie apocalypse is the nearest shopping mall. Between the food court, the sporting goods store, and the toy shop, you will find everything you need comfortably to wait out the mindless cannibal hordes, the barbarians at the gate who, like you, only want to survive, though they can no longer remember why. From here, safely ensconced behind the great walls of the shopping mall with your creature comforts and commodities, you might be able to rebuild the world. You might even make America great again.

At bottom, these are two variations on one theme. The messianic-redemptive takes place on the edge of Armageddon, while the naturalistic-sentimental is set just on the other side of it. The audience rocks to and fro across the divide between them, like the patrons in Adams' impossible restaurant at the End of the Universe.[15] Capitalism thereby ensures its futurity by projecting the present onto a fabricated future of its own making and under its control, numbing us to the (relatively) slow environmental conflagration we are already living and, as with any opiate, turning a profit by doing so. We are thereby enjoined to enjoy the illusion that even after the great catastrophe, the world will go on just as it had before, a little worse for the wear but not really different for all that. The Hollywood fixation on the end of the world is, in fact, anything but; to the contrary, it evinces a nostalgic yearning for a simpler time that, as with all objects of nostalgia, never really was.

We should put a finer point on all this. The trouble here is not simply that the catastrophic imaginary breeds a complacency that prevents our acting on behalf of our world before it is too late. The trouble is, all these variations on the end of the world imagine precisely the opposite—the preservation and ennobling of the world's existing terms and conditions, the repeated reification of the established parameters of possibility—because *it is already too late*. We watch the world explode, maybe dress up for it now and then to give it a sense of occasion, because, as a matter of fact, the world ended some time ago and we are merely living on through the slow process of ruination, the unspectacular and even boring catastrophe, we call modernity. More troubling still, the world may never have existed in the first place.

A Psychotic Hypothesis: From the Museum to the Mausoleum

The world is over, the world did not exist.

Such an assertion may seem to walk the path already tread by the psychotic Dr. Schreber who, having watched the world disintegrate before his eyes, was forced to create it anew through the laborious fabrication of his own delusion.[16] And why not? Lacan insisted that the truth at stake in the psychotic's testimony is universal insofar as it exposes what the world, any world, truly is: a question of experience and its relation to the linguistic mediation through which that experience is made to signify. The world is a matter of interpretation or it is literally meaningless. The difference between Schreber's world and ours thus is not qualitative but dispositional. It is not that the psychotic is wrong and the rest of us are right; if anything, Schreber is the more right to recognise the cataclysmic defect in the Order of the World, the hole in the symbolic, to which his being—or, *qua* Lacan, the being of any subject—corresponds. The psychotic's lament and the source of his persecution is that he is unbearably alone with his reality, which is no more or less real than the reality the rest of us supposedly share. "Reality," then, merely names our common means of disavowing the defect in the symbolic, the lack in the Other, that Schreber was only too right to recognise as cataclysmic.

To collapse the parallelism which, Baudrillard reminds us, holds the future infinitely at bay in the face of all facts, we therefore should take a page from the psychotic's book. Let us admit that the world is over, that capitalism is already the catastrophe we have been waiting for. The capitalist interpretation was only ever that, an interpretation. And one, moreover, whose symbolic coordinates have been exhausted, even as they insist ever more emphatically upon their perpetuity. Faced with such a truth, our question is also the psychotic's question, from which the new delusion unfolds: *Now what?*

No amount of reality-checking or rational entreaty against those ever more desperate or absurd efforts to extend the world's capacities for suicidal excess will save it, for all such efforts cannot but operate within the same parameters and according to the same symbolic coordinates of that absurd desperation. What is more, they embellish the very fiction—that is to say, they reify the very reality—they mean to arrest, because the endless stream of calamitous news about the biosphere and the litany of warnings that we must act quickly to avoid the worst cannot escape entrapment within the catastrophic imaginary which frames and encourages them. After Baudrillard, we know there is no outside the simulation. More facts, more evidence, ever more dire predictions about the end-to-come—such rational appeals can fool us, as they may fool the well-meaning appellants themselves, into believing that the end is nigh, when in fact the end is now.[17] As Frank Ruda has argued *apropos* of his call for a "comic fatalism," the worst has already happened; our task, and this was also Schreber's task, is to think the worstness of the worst.[18]

Ruda's comic addendum may be novel, but his fatalism is not. Alongside T. S. Eliot, who most famously noted the whimper with which the world ends, consider that other great prophet and poet of modernity, Claude Lévi-Strauss, particularly his 1955 memoir *Tristes Tropiques*, a text that in fact is less memoir than work of mourning. No matter how far and wide his travels or the purported exoticism of his informants, inevitably the great anthropologist finds only the ruins of what had been or, at best, cultures irreversibly corrupted by what is called progress, which can itself move in but one direction. He finds not cultures but the living corpses of culture, animated and puppeteered by their own slow death, driven towards the inertia that results from the elimination of all difference. Such a flattening of being is patently the world-ordering, world-making ambition of the civilisation from which he orients his own outward-facing gaze, through which he discovers, everywhere he looks, only so much wreckage of this urge to sameness which is the hideous reflection of himself and his world.[19] A strange conclusion, perhaps, for an anthropologist, whose work therefore is better suited to the mausoleum than the museum, who therefore might better be termed an *entropologist*. But this dolorous rigour imposes upon us the necessity—ethical in kind, and consistent with Ruda's injunction as well as our wariness towards the lure of nostalgia—not of preserving what remains but of mourning what has been lost and can never be again.

If we thus know the world has already ended, maybe ended before it began, then to refuse Lévi-Strauss' invitation to mourn, to instead repeat and repeat the imaginary end-to-come, is an instance of melancholia which, as such, is immune to empirical correction because its domain is not reality but fantasy. As is well known, Freud's account of the melancholic dilemma holds that it instances an apparent masochism which is actually its opposite: a sadism towards some lost object that had been invested with a quantity of libido that recoils onto the subject when it is deprived of its place-holder in the imaginary. It is there in the mask of masochism where the subject takes her revenge upon the object. In our collective case, the object is the world we still desire but can no longer (and, in fact, could never) have. But this displaced sadism is more than a pleasure in cruelty; it is this, even if it is the sadist who suffers, but its full function is to retain the illusion of our power over the world, at once acknowledging and denying the very reality upon which we impose our vengeance. The apocalypse is a lover's discourse.

This further reveals the infantilism at play even in the alarm bells singing constantly that *if* civilisation does not make some immediate, massive changes to its industrial model, infrastructure, and public values, the bad news about the environment will only get worse. The trouble with mourning—its challenge but also its promise, such as it is—is that it refuses the optimism implied by the "if" in the previous sentence. Against the temptation to endorse a sentimental optimism, however faint, we should hold to Lévi-Strauss' invective against the high priests of progress: the death drive which is pushing civilisation as we know it to the point of total self-cancellation, even as this movement is disavowed or repressed through ever-more elaborate exercises of the pleasure principle, passed its point of no return a long time ago. Or better, the impulse to cancel itself out is a feature of civilisation, not a bug. It is encoded into the structure of modernity, as both its object and cause. The optimist's interpretation of the alarm, meanwhile, still clings to the same fantasy of mastery at the root of the problem, the fantasy that places the human over and against its environment, investing it with the same cannibalistic appetite that has consumed the environment beyond revival—at least beyond any revival of what it once was or was supposed to be. More alarming still, the demand that we pull the emergency brake on the planet's material depredation cannot but participate in and perpetuate the catastrophic imaginary, and therefore is attended by the same perverse enjoyment that makes the apocalypse so profitable, both as a subject of entertainment and as a necessary fact of industrial capitalism.

To find our way beyond this symptomatic discourse and the impasse with which it confronts us, psychoanalysis can help inform a shift from melancholia to mourning, unbinding the knot joining the catastrophic imaginary to the pleasure principle. It can help us think the question of responsibility, of *Now what?*, not in terms of our obligation to preserve the world, or as much of it as can be salvaged, for the future, but rather with regard to futurity as such. As Alenka Zupančič has written in response to Ruda's comic fatalism, "The world will surely end, but it won't be the end of our troubles."[20] To shift from melancholia to mourning is to dissolve the libidinal bonds enchaining us to a past that never really was, opening us to a future that is informed but not determined by the world we will have left behind. For psychoanalysis is, more than anything, about life after death, about the death we are always already living, about how to live with rather than against—to live *through*—the death which drives us.

This is not a metaphysical or religious hypothesis. The point, rather, is that the world is not the planet. Life goes on after the end of the world. We therefore have to consider, like Schreber, how to live not in but *after* the end times, beyond the world which has ended, to let it go and move towards another, new world, the parameters of which have yet to be established and, indeed, are incalculable from within the imaginary and symbolic limits of the melancholic present. Precious though it may be, we have to leave this world behind, to have done with it as Antonin Artaud asked us to have done with the judgment of God.[21]

To have done with the world thus is not to accept its ending with the shrug of the passive nihilist, any more than Schreber could accept the cataclysm only he could remediate. The ground of psychoanalysis, as Lacan repeatedly makes clear, is an ethical concern for the subject's becoming, against the coded expectations, limitations, ways of knowing—in short, against the very language, with all it signifies—that banish each subject's singular humanity to the other side of the social link. There is therefore both a self-regarding and an outward-facing aspect to the ethics of psychoanalysis: the steadfast journey towards self-knowledge along the royal road of the unconscious presumes, even requires, that the reality in which the unconscious is repressed, the symbolic strictures obscuring the path of truth, be acted upon and transformed by the subject as she makes a way for her desire. Psychoanalysis therefore invites us to follow Lévi-Strauss into the mausoleum, and there to undertake the work of mourning through which we might loosen our libidinal attachments to both the catastrophic imaginary and its inverse, the real catastrophe we are living and after which we will have to go on. We should acknowledge that the end of the world is boring, even mundane, that the slow-moving catastrophe of which we are a part is irreversible. And we must admit that neither a science nor a God will save us. Easier said than done, surely. No doubt painful and terrifying. But this is why the question prior to any possible treatment of the everyday catastrophe that surrounds us and in which we, too, play our own parts is the question of ethics. Freud once called psychoanalysis a "school of suffering."[22] Let the lesson commence.

Notes

1. Renee Lertzman, *Environmental Melancholia: Psychoanalytic Dimensions of Engagement* (New York: Routledge, 2015), xiv, 151.
2. Douglas Adams, *The Restaurant at the End of the Universe*, repr. ed. (New York: Del Ray, 1995), 100–104.
3. See, for instance, Fredric Jameson, *An American Utopia: Dual Power and the Universal Army*, ed. Slavoj Žižek (New York: Verso, 2016), 3.
4. Louis Althusser, *On the Reproduction of Capitalism: Ideology and Ideological State Apparatuses*, trans. G. M. Goshgarian (New York: Verso, 2014), esp. 232–70.
5. Jacques Lacan, *The Seminar of Jacques Lacan, Book XX: On Feminine Sexuality: The Limits of Love and Knowledge, Encore, 1972–1973*, ed. Jacques-Alain Miller, trans. Bruce Fink (New York: W. W. Norton, 1998), 1.
6. See Jean Baudrillard, "Global Debt and Parallel Universe," trans. Francois Debrix, *Event-Scenes* (October 16, 1996); http://ctheory.net/ctheory_wp/global-debt-and-parallel-universe/ (accessed January 2020); also, Nathan Gorelick, "The Real (of) Debt: Notes Toward an Ethics of Trash," *Continental Thought and Theory* 1, no. 2 (January 2017): 491–517, https://ir.canterbury.ac.nz/handle/10092/13088 (accessed January 2020).
7. For an extended discussion of this point, see Nathan Gorelick, "Why Sex is Special: Psychoanalysis against New Materialism," in *Subject Lessons: Hegel, Lacan, and the Future of Materialism*, ed. Russell Sbriglia and Slavoj Žižek (Evanston, IL: Northwestern University Press, 2020), 171–89.
8. See films: Jonathan Hensleigh and J. J. Abrams, *Armageddon*, directed by Michael Bay (Burbank, CA: Touchstone Pictures, 1998); Roland Emmerich and Dean Devlin, *Independence Day*, directed by Roland

Emmerich (Los Angeles: Twentieth Century Fox, 1996); Luc Besson and Robert Mark Kamen, *The Fifth Element*, directed by Luc Besson (Culver City, CA: Columbia Pictures, 1997); Roland Emmerich and Jeffrey Nachmanoff, *The Day After Tomorrow*, directed by Roland Emmerich (Los Angeles: Twentieth Century Fox, 2004); Roland Emmerich and Harald Kloser, *2012*, directed by Roland Emmerich (Los Angeles: Columbia Pictures, 2009); Bruce Joel Rubin and Michael Tolkin, *Deep Impact*, directed by Mimi Leder (Hollywood, CA: Paramount Pictures, 1998); and Dean Devlin and Paul Guyot, *Geostorm*, directed by Dean Devlin (Burbank, CA: Warner Bros., 2017).

9. Daniel Defoe, *Robinson Crusoe*, ed. Michael Shinagel (New York: W. W. Norton, 1993).
10. See films: Alex Garland, *28 Days Later*, directed by Danny Boyle (London: DNA Films, 2002); Cormac McCarthy and Joe Penhall, *The Road*, directed by John Hillcoat (New York: Dimension Films, 2009); George Miller and James McCausland, *Mad Max*, directed by George Miller (Potts Point, Australia: Kennedy Miller Productions, 1979); Peter Rader and David Twohy, *Waterworld*, directed by Kevin Reynolds (Universal City, CA: Universal Pictures, 1996); Andrew Stanton, Pete Docter, and Jim Reardon, *WALL-E*, directed by Andrew Stanton (Burbank, CA: Walt Disney Pictures, 2008); and Lana Wachowski and Lilly Wachowski, *The Matrix*, directed by Lana Wachowski and Lilly Wachowski (Burbank, CA: Warner Bros., 1999).
11. Sigmund Freud, *Moses and Monotheism*, in *The Standard Edition of the Complete Works of Sigmund Freud*, ed. James Strachey et al. (London: Hogarth, 1955), vol. 23, esp. 80–92; and *Civilization and Its Discontents* in *The Standard Edition*, vol. 21, 74.
12. Freud, *Civilization and Its Discontents*, in *The Standard Edition*, vol. 21, 74.
13. Freud, "Creative Writers and Day-Dreaming," in *The Standard Edition*, vol. 9, 149–50.
14. Friedrich Nietzsche, *The Gay Science*, trans. Josefine Nauckhoff (Cambridge, UK: Cambridge University Press, 2001), 120.
15. This rocking back and forth across the threshold of apocalypse further recalls the child's game of "*fort/da*" Freud examined on his way to theorising the death drive. See Freud, *Beyond the Pleasure Principle*, in *The Standard Edition*, vol. 18, 12–17.

16. Daniel Paul Schreber, *Memoirs of my Nervous Illness*, trans. Ida Macalpine and Richard A. Hunter (New York: New York Review Books, 2000).
17. This is a matter of material fact. See Jem Bendell, "Deep Adaptation: A Map for Navigating Climate Tragedy," *IFLAS Occasional Paper* (July 27, 2018): 1–36, https://www.lifeworth.com/deepadaptation.pdf (accessed January, 2020). If Bendell's insights are new to you, apologies in advance for ruining your day.
18. Frank Ruda, *Abolishing Freedom: A Plea for a Contemporary Use of Fatalism* (Lincoln: University of Nebraska Press, 2016).
19. Claude Lévi-Strauss, *Tristes Tropiques*, trans. John Weightman and Doreen Weightman (New York: Penguin, 2012), esp. 37–38, 43–44, 321–415 (on anthropology vs. entropology, see 413–14).
20. Alenka Zupančič, "The End," *Provocations Journal* 1 (October 31, 2016): 1–9 https://www.provocationsbooks.com/2016/10/31/the-end/ (accessed January 2020).
21. Antonin Artaud, "*To Have Done with the Judgment of God*, a radio play (1947)," in *Antonin Artaud: Selected Writings*, ed. Susan Sontag, trans. Helen Weaver (Berkeley: University of California Press, 1988), 555–71.
22. Freud, *Notes Upon a Case of Obsessional Neurosis*, in *The Standard Edition*, vol. 10, 209.

Filmography

Besson, Luc, and Robert Mark Kamen. 1997. *The Fifth Element*. Directed by Luc Besson. Culver City, CA: Columbia Pictures.

Devlin, Dean, and Paul Guyot. 2017. *Geostorm*. Directed by Dean Devlin. Burbank, CA: Warner Bros.

Emmerich, Roland, and Dean Devlin. 1996. *Independence Day*. Directed by Roland Emmerich; Los Angeles: Twentieth Century Fox.

Emmerich, Roland, and Harald Kloser. 2009. *2012*. Directed by Roland Emmerich. Los Angeles: Columbia Pictures.

Emmerich, Roland, and Jeffrey Nachmanoff. 2004. *The Day After Tomorrow*. Directed by Roland Emmerich. Los Angeles: Twentieth Century Fox.

Garland, Alex. 2002. *28 Days Later*. Directed by Danny Boyle. London: DNA Films.

Hensleigh, Jonathan, and J. J. Abrams. 1998. *Armageddon*. Directed by Michael Bay. Burbank, CA: Touchstone Pictures.

McCarthy, Cormac, and Joe Penhall. 2009. *The Road*. Directed by John Hillcoat. New York: Dimension Films.
Miller, George, and James McCausland. 1979. *Mad Max*. Directed by George Miller. Potts Point, Australia: Kennedy Miller Productions.
Rader, Peter, and David Twohy. 1996. *Waterworld*. Directed by Kevin Reynolds. Universal City, CA: Universal Pictures.
Rubin, Bruce Joel, and Michael Tolkin. 1998. *Deep Impact*. Directed by Mimi Leder. Hollywood, CA: Paramount Pictures.
Stanton, Andrew, Pete Docter, and Jim Reardon. 2008. *WALL-E*. Directed by Andrew Stanton. Burbank, CA: Walt Disney Pictures.
Wachowski, Lana, and Lilly Wachowski. 1999. *The Matrix*. Directed by Lana Wachowski and Lilly Wachowski. Burbank, CA: Warner Bros.

Bibliography

Adams, Douglas. 1995. *The Restaurant at the End of the Universe*. Reprint ed. New York: Del Ray.
Althusser, Louis. 2014. *On the Reproduction of Capitalism: Ideology and Ideological State Apparatuses*. Translated by G. M. Goshgarian. New York: Verso.
Artaud, Antonin. 1988. *To Have Done with the Judgment of God*, a radio play (1947). In *Antonin Artaud: Selected Writings*. Edited by Susan Sontag. Translated by Helen Weaver, 555–71. Berkeley: University of California Press.
Baudrillard, Jean. 1996. Global Debt and Parallel Universe. Translated by Francois Debrix. *Event-Scenes*. October 16. Accessed March 2021. https://journals.uvic.ca/index.php/ctheory/article/view/14840/5711.
Bendell, Jem. 2020. Deep Adaptation: A Map for Navigating Climate Tragedy. *IFLAS Occasional Paper*, Revised 2nd Edition (July 27): 1–35. Accessed March 2021. https://jembendell.com/2019/05/15/deep-adaptation-versions/.
Defoe, Daniel. 1993. *Robinson Crusoe*. Edited by Michael Shinagel. New York: W. W. Norton.
Freud, Sigmund. 1953–1966. *The Standard Edition of the Complete Works of Sigmund Freud*. Edited by James Strachey et al. 24 vols. London: Hogarth.
Gorelick, Nathan. 2017. The Real (of) Debt: Notes Toward an Ethics of Trash. *Continental Thought and Theory* 1 (2): 491–517, January. Accessed March 2021. https://ir.canterbury.ac.nz/handle/10092/13088.
———. 2020. Why Sex is Special: Psychoanalysis against the New Materialism. In *Subject Lessons: Hegel, Lacan, and the Future of Materialism*, ed. Russell

Sbriglia and Slavoj Žižek, 171–189. Evanston, IL: Northwestern University Press.
Jameson, Fredric. 2016. *An American Utopia: Dual Power and the Universal Army*. Edited by Slavoj Žižek. New York: Verso.
Lacan, Jacques. 1998. *The Seminar of Jacques Lacan, Book XX: On Feminine Sexuality: The Limits of Love and Knowledge, Encore, 1972–1973*. Edited by Jacques-Alain Miller. Translated by Bruce Fink. New York: W. W. Norton.
Lerztman, Renee. 2015. *Environmental Melancholia: Psychoanalytic Dimensions of Engagement*. New York: Routledge.
Lévi-Strauss, Claude. 2012. *Tristes Tropiques*. Translated by John Weightman and Doreen Weightman. New York: Penguin.
Nietzsche, Friedrich. 2001. *The Gay Science*. Translated by Josefine Nauckhoff. Cambridge, UK: Cambridge University Press.
Ruda, Frank. 2016. *Abolishing Freedom: A Plea for a Contemporary Use of Fatalism*. Lincoln: University of Nebraska Press.
Schreber, Daniel Paul. 2000. *Memoirs of my Nervous Illness*. Translated by Ida Macalpine and Richard A. Hunter. New York: New York Review Books.
Zupančič, Alenka. 2016. The End. *Provocations Journal* 1: 1–9, October 31. Accessed March 2021. https://www.provocationsbooks.com/2016/10/31/the-end/.

13

From the Sublime to the Hysterical Sublime: Reading the End of the World Against the Singularity

Matthew Flisfeder

Classical liberal philosophy, like Thomas Hobbes' *Leviathan*, justifies the political order through a of mythical "state of nature," prior to civilisation. Hobbes describes the state of nature as a war of "every man against every man." His solution is that individuals should give up their rights to self-defence and place such authority into the hands of the Sovereign to whom all submit as the only *legitimate* force. Inconsistencies of (human) nature revert here, as they still do today, into justifications for the reigning order. Neoliberals, for instance, argue for the naturalness of market competition with reference to Hobbes' state of nature. It is interesting, then, that authors like Geoff Mann and Joel Wainwright return to *Leviathan* to theorise the political conditions of the *contemporary* state of nature, that is, of climate change. In *Climate Leviathan*, Mann and Wainwright examine the crisis of climate change according to two intersecting political and economic problematics: that of capitalism and that of the state.[1] They ask whether or not the world to come in the face of

M. Flisfeder (✉)
The University of Winnipeg, Winnipeg, MB, Canada
e-mail: m.flisfeder@uwinnipeg.ca

C. Burnham, P. Kingsbury (eds.), *Lacan and the Environment*, The Palgrave Lacan Series, https://doi.org/10.1007/978-3-030-67205-8_13

climate change will continue to be capitalist; and, whether or not the world will be governed by some kind of planetary Leviathan-like sovereign to mitigate the effects of the crisis. The intersection between these problematics results in four possibilities that inform their matrix of analysis: they call the capitalist sovereign the Climate Leviathan; the capitalist non-sovereign they call the Climate Behemoth; the non-capitalist sovereign they call Climate Mao; and, the non-capitalist and non-sovereign they call Climate X. Climate Leviathan is somewhat akin to Hobbes' Leviathan, where capitalism and the capitalist class state persists into the future. Climate Behemoth, in contrast, appears to be something like a reactionary formation within capitalism. Climate Mao correlates with the Chinese model of state-authoritarianism; whereas, Climate X proposes the as yet unknown anti-state and anti-capitalist alternative.

The models that Mann and Wainwright devise are similar to those proposed by Peter Frase in his book, *Four Futures*, where he suggests that the twin problems of the twenty-first century will be climate change and automation.[2] Climate change and automation will force us to contend with questions about scarcity and inequality. Frase develops a political schema ("four futures") for what may result. A world of equality and abundance he calls Communism; the world of hierarchy and abundance, he calls Rentism; the world of equality and scarcity, he calls Socialism; and, the world of hierarchy and scarcity, he calls Exterminism. The four futures overlap with the climate Leviathan, although with some differences. Notably, Frase considers the climate crisis measured against the dilemma of automation. Doing so, I agree, allows us to more properly measure the human ethical dimension, caught between nature and culture, or technology. Two "ends" of sorts structure this dimension: the end of nature and the end of humanity—the "end of the world" or the so-called singularity. On this point, I believe psychoanalysis provides an ethical program for conceiving this dilemma, which I approach through the aesthetic category of the sublime.

How the Signifier Falls Into the Signified

Sigmund Freud writes in *Civilization and Its Discontents* that, "as a member of the human community, and with the help of technology guided by science, one can go over to the attack on nature and subject it to the human will."[3] The implications in this statement resonate well with a Hegelian universalism that departs from the kind of Spinozist and Deleuzian materialism occupying much critical theory today (particularly in the various new vitalist materialisms, object-oriented ontologies, and posthumanisms). If we consider the Hegelian proposal that the subject emerges as a crack in the purely immanent substance of the world (that is, not only as substance but also as subject), then we can also see to some degree how the freedom that emerges from the negation of the immediate (through communication, through sexual *perversions*, through technology, etc.) also brings forth the kind of alienation that Freud addresses in *Civilization and its Discontents*. Immediacy, as Gregor Moder puts it, is always a lost immediacy—it is an immediacy that we can only know from the position of alienation.[4] Such an immediacy, lost prior to (or as a result of) subjectivisations, is only apprehensible in the form of the fantasy, which remains as the source of our sensation of alienation, which after all is what propels us into the position of desiring, and which is our real source of enjoyment. By subjecting nature to the human will, an alienation—a discontent—thus is born in the human community, which is at the same time the universal condition of the desiring subject.

The view expressed in Freud's statement contains a number of ramifications with regards to both the analytical discourse, as well as for the relationship between the human subject and the non-human. Psychoanalysis shares with historical materialism the insight that knowledge bears upon the subject's acting on both the world and itself, and is therefore capable of transforming both. This is one way of interpreting the Lacanian logic of the signifier falling into the signified, or of the Symbolic having an effect in the Real.[5] Discourse reifies abstract phenomena into objects of knowledge allowing us to act upon and so change them. Objects do not exist merely in discourse, but their representation in discourse allows us to so transform them, producing an equal transformation in both substance and subject. The analytical discourse is the product of the hysterical subject who *creates* new knowledge: the foundation upon which analysis is based. The hysteric troubles the Other with the question: *ché vuoi?* What am I to you? Why am I what you (*le grand Autre*) are saying that I am? It is a question that so begins the production of the analytical knowledge/discourse.

The scientific discourse operates in this way, too. It translates empirical data into objects upon which we can act. Nature, for instance, is an abstract phenomenon—Real, material, but nonetheless abstract at the level of our cognition—it is *das Ding*. Science makes it possible to extract from nature, to appropriate, and create the new. This is how the scientific discourse develops new technologies. The creation of the new has the reciprocal dimension of impacting upon us, on our know-how, on the evolution of the human subject, which is what gives knowledge its historical fervour. The fact that knowledge makes possible our acting upon the world indicates the utter necessity of ethical thinking given that our individual and particular actions have universal consequences: not only consequences for ourselves, but for the entirety of the human community, impacted as it is by the non-human—that is, by the substance that is always already lost. *Ethics*, that is, not as some kind of autonomous or normative realm, separate from the social reality; ethics is the product of thinking, of reasoning. The act, after all, is the result of taking thought all the way to its reasonable ends.

Psychoanalysis is much more far-reaching and universal than we might expect here. The ethics of psychoanalysis is a universalism insofar as the subject is interpellated at a foundational level by the Symbolic order, which precedes it. The new is produced when the subject brings to consciousness, through its own ethical gesture of not giving way to its desire (that is, the series of negations—"that's not it"—that drive the subject's enjoyment), that *it is itself the positing source of its own limitations*. But there is a sense in which the completion of the analytical treatment is only fully possible in conditions of universal emancipation, where the conscious *apprehension* of alienation becomes universal. In this sense, there is a correlation between the ethics of psychoanalysis and historical materialism. The two overlap around the ethical ambiguities of two interrelated phenomena today: the existential threat of climate change, and the rise of new technological advancements in machine learning.

These phenomena are interrelated at the level of production in as much as the growth of technology has historically been the result of our extractions from nature, based upon certain historical stages in our scientific knowledge, as well as the various historical stages in the mode of production, which has led to developments in land cultivation and agriculture, industrialisation, mechanisation, and digital automation. The age of the Anthropocene, thus, forces us to ask, what is to be done?

This dilemma is conflated by two morbid or gothic fantasies of the present age: the fantasy of the end of the world, and that of the technological singularity. Each in its own way is a fantasy of a return to immediacy and the end of humanity as we know it. The former is best rendered by Fredric Jameson's thesis: "it seems easier for us today to imagine the thoroughgoing deterioration of the earth and of nature than the breakdown of later capitalism."[6] I read this statement against the Autonomist Marxist reading of the "general intellect" found in Marx's "Fragment on Machines" in the *Grundrisse*.[7] General intellect is a description that comes close to the notion of the technological singularity surpassing human intelligence. These two fantasies relate to the two sides of the modern and the postmodern sublime: the natural and the technological.

The aesthetic category of the sublime has been read more traditionally, within the realm of nature or the natural; but there is also a sense here of the *hysterical* sublime that Jameson identifies in his writing on postmodernism. There is an hysterical sublime discovered in the mechanisms humanity has devised in the process of self-alienation from nature, and from our attempts to so dominate it. But this sublime is of the second order, whereby our universal admiration and inspiration from nature has been so transformed into a representation of the deleterious, perhaps even threatening dimensions of human culture, producing out of nature our technologies, which may now come back to bite us in the ass. The sublime object, and the hysterical sublime, thus offer us an avenue for conceiving this dilemma in the Lacanian-Hegelian register.

The Most Sublime Hysteric

The *experience* of the sublime for Georg Wilhelm Friedrich Hegel is similar to that described by Immanuel Kant. According to Kant, the sublime can be thought in relation to representation, and develops out of the distinction between phenomena and noumena. Phenomena are those experiences that can be easily translated and represented in discourse that make them comprehensible and more concrete. Noumena, however, refer to those objects that stand outside of human sense perception, which we are not able to apprehend due to the limitations of our minds and bodies. For Kant, we can know only our knowledge of things, but we cannot know things-in-themselves. The sublime, then, for Kant, speaks to our ability to come closer to knowing the thing-in-itself, even if we can never breach that limitation. The difference for Hegel lies with the way that he (as a reaction to Spinoza) *subtracts* the idea that some substantial *Thing* persists *beyond* our experience.

For Hegel, our experience of contradiction is what indicates the rupture of pure immanence. For him, the sublime brings us close to the gap in phenomenal reality. Hegel's position is that the "negative experience of the Thing must change into the experience of the Thing-in-itself as radical negativity."[8] What Kant perceives as a limitation to our experiences of reality is for Hegel the very reality itself. It's not that we are limited from experiencing the true substance of reality; for Hegel, the limit just *is* reality. Reality is itself traversed by a series of cracks and fissures. What lacks in the subject is redoubled into substance *as* lacking, or as always already inconsistent with itself. For Hegel, the way to overcome the limits of representation is by recognising that immediacy is always a lost immediacy. Absolute Knowing is rendered by the teleological within the sublime. The sublime allows us to apprehend the Absolute in its failure to be represented adequately.[9] I call this experience the "hysterical sublime" or the "hysteric's sublime." It is the sublime as conceived by Hegel, who Jacques Lacan referred to as "the most sublime hysteric" (*le plus sublime des hysteriques*)[10]—thus, the hysteric's sublime; but it is also the sublime of the hysteric who perceives the crack in the Other, and by questioning it, produces new knowledge. I'll elaborate.

We need to begin with the *origin story* of the subject in the forced choice of being. As Slavoj Žižek recalls, Lacan develops this line in two different places where he divides the Cartesian *cogito* into the "I think" and the "I am." In Seminar XI, *The Four Fundamental Concepts of Psycho-Analysis*, Lacan claims that the subject chooses thought and loses being.[11] However, a few years later, in Seminar XIV, *The Logic of Fantasy*, he proposes instead that the subject is forced to choose being, and thought is relegated to the position of the unconscious: I am, therefore it (the unconscious) thinks. Rather than read these two versions as mutually exclusive, Žižek proposes interpreting them according to the logics of sexuation. The masculine subject opts for the second choice, that of being; whereas the feminine subject chooses thought.[12] We can justify this distinction when we read the two logics according to their orientation towards the phallic signifier; and we need to read this orientation according to the binary logic of the affirmation and the negation. Whereas the masculine subject is the one who affirms the phallus as the signifier of castration (choosing being over thought), the feminine subject negates the phallic signifier; however, she is still oriented towards it in a relation of negation (choosing thought over being). Masculine and feminine are not *biological* identities; they are identifications with the signifier. The masculine subject affirms it, whereas the feminine subject identifies with it in the form of the questioning or the troubling, that is, the thinking, of it. This fact connects to our thinking of the logic of the fantasy, and its relation to the sublime object.

In the forced choice of being, the subject is caught in a primordial act of affirmation and negation. By affirming the signifier ("I am this")—which reflects how the subject perceives itself from the perspective of the Other, of the Symbolic order—the subject negates what it is not through what it perceives as the Other's desire. But in this very act of negation, there is a remainder, a fantasy of the choices not chosen, and this is what produces the subject as desiring. We come to understand the Lacanian *object a* as the very objectification of this lack—the lack identified as the choices not chosen. The limit here, for the subject, is unconsciously self-imposed (it and not the Other makes the choice), but in order to perceive the object as phenomenally possible it must displace this obstacle onto the Other. While we equate desire with the lost object, what we truly desire is the obstacle, the limit, without which we would realise that the object does not exist.[13] We think that enjoyment will derive from attaining the lost object—lost in our very act of choice, of negating it—when in fact, we always already procure enjoyment from the fantasy.[14] This lost object is what is sublime in the Lacanian register, and reflects the lost immediacy of a foundational substance. It is a remainder/reminder of what we negate to become what we are. This negation demonstrates in the Hegelian register that, just as the subject is lacking, so too is substance.

Put differently, and in terms that are relevant for *hystericising* substance, we must recognise that no possibility exists for creating some new harmony between humanity and nature. This is where some of the vitalist new materialists fail, by perceiving the potentials for a renewed harmony simply by withdrawing human agency, what Steven Shaviro, for instance, theorises as an escape from Anthropocentrism through a strategic Anthropomorphism.[15] For me, this fantasy for a renewed harmony—a return to lost substance—is one of the central fantasies of the Anthropocene. We desire the *fantasy* of a lost nature more so than the possibility of a renewed harmony; and, its inverse is not a fantasy of a renewed harmony, but the fantasy of the end of the world—of total destruction from the climate disaster.

From The Hysteric's Sublime ...

The hysteric and feminine subject overlap in their position of questioning. Whereas the masculine subject, in affirming the phallic signifier, adopts the identity it registers in the Other's desire (I am this), the feminine subject is in a place of questioning the Other: why am I what you (the big Other) are saying that I am? This question is the one, still, posed by all subjects at that foundational point of the primordial forced choice, and the resulting choice is an attempt to satisfy the desire of the Other. But in the case of the feminine hysteric, the question is turned back to the Other asking why am I what you are saying that I am? We should read Lacan's discourse of the hysteric in precisely these terms.

In the formula for the hysteric's discourse, the subject ($) is in the position of the agent, the master-signifier (S_1) is in the position of the worker, knowledge (the Symbolic) is in the position of production, and the *object a* is in the position of truth.

$$\frac{\$}{a} \rightarrow \frac{S_1}{S_2}$$

By questioning the master-signifier—*Chè vuoi?*—the hysterical subject is responsible for helping to procure a new knowledge (S_2). Specifically, in this context, the hysteric is what drives the production of the Freudian analytical discourse. The hysteric sets it in motion and allows the analytic discourse to formulate its own object: the unconscious. This new knowledge reveals that coming to grasp the object *as* an object of discourse is the only way to create the conditions for so acting upon it and transforming it (and consequently the subject, too). The idea that knowledge changes reality, as Žižek notes, is what psychoanalysis shares both with scientific discourse as well as with historical materialism.[16]

If the masculine enjoyment is the enjoyment of the fantasy ($\$<>a$), then the feminine enjoyment, according to Lacan, is tied to the Symbolic, to the production of knowledge (or even knowing, that is, absolute knowing in the Hegelian sense—the capacity to reason, to experience the infinite and unlimited). But there is a corollary here, as well, with the production of scientific knowledge and with the scientific discourse. We see this manifestly in the production of embodied knowledge in the form of technology—what Lacan playfully calls *lathouse*.[17] *Lathouse* are, as Žižek explains, embodied scientific knowledge (a kind of surplus-knowledge).[18] They are what Clint Burnham describes as surplus objects embodying *jouissance*, which did not precede the subject in some primordial substance, like nature. They are technological prosthetics that embody libidinal enjoyment, but which have been developed by human subjects.[19] This is what calls forth the need to have ready the theory of the postmodern hysterical sublime that Jameson describes.

… To the Hysterical Sublime

For Jameson, the postmodern hysterical sublime is distinguished from the modern sublime for the fact that it reveals our relationships with technology rather than with nature. It contains still those feelings of amazement, awe, and astonishment that we find in the modern conception of the sublime (even the teleological and the ethical), only now what shocks us are the very products of our own invention: the technological (and even cultural, that is, to cultivate—the human act of so transforming nature). Just as Freud describes how the human community may subject nature to the human will through technology, now we fear most losing control of the very thing that we ourselves have created. There is a sense in which a conception of the hysterical sublime was also found in Marx, particularly in his rendering of the "general intellect."

In the *Grundrisse*, Marx writes that the development of machinery is the culmination of the different metamorphoses through which the labour process undergoes in the production process of capital. It can or will at some point in its development transform into a system of automatic production, put in motion by a power that moves itself.[20] Machinery, he writes, objectifies and appropriates living (human) labour, coming to dominate it in the very form of capital. Automation is the complete objectification, not only of living labour, but of the general intellect—that is, of the accumulation of society's scientific knowledge.[21] Two opposed fantasies present themselves with the contemporary realisation of Marx's reading of automation: first, is the accelerationist fantasy of a world without work; the second is the hysterical sublime, the nightmare fantasy of a world without *us*.

Accelerationist thinkers, like Nick Srnicek and Alex Williams, foresee the possibility of a fully automated society (what some refer to as fully automated luxury communism), where the drudgery of labour is alleviated due to the growth of labour saving technology.[22] However, given the historical conditions and contexts of capitalist society, many others now fear the steady replacement and stagnation of labourers as a cost saving practice of capital (that is, "finding efficiencies"), which as Jameson argues is the true plot of Marx's *Capital*.[23] It is, he claims, the story of unemployment as an inevitable course for capitalism, rather than merely class struggle. Of course, greater proletarianisation is one of the central causes of capitalist crisis, when wage labour is the source of surplus value in the production process. For capitalists to maintain the production of surplus value in a fully automated society they would have to pay the machines for their labour (which would, of course, contradict the very reason for their employment in the first place). From my perspective, it is the fact of proletarianisation that outweighs the rise of the general intellect—or the "singularity," the hypothetical future moment when the growth of technology becomes so uncontrollable and inevitably overpowers the human civilisation—turning around Freud's remark that the human will overpowers nature through technology; now technology overpowers us since it is the stagnation of human subjects that make such a transformation political.

Jameson, for his part, theorises the hysterical sublime as the manner in which the whole, or the totality, of the capitalist mode of production is experienced in postmodern capitalism. What we come to grasp through the hysterical sublime is not merely the fear and panic set up in the face of the singularity, but of the total and maybe even unrepresentable dimensions and reach of the entire network of global capitalism. Technology becomes a way to perform the kind of cognitive mapping that Jameson proposes as a means to radically transform our conditions in late capitalism; and, it is important that he describes this process, drawing on Lacan, as the product of the Symbolic having an effect in the Real.[24]

Capitalism has proven itself to be a force that has so gone on the attack on nature, to exploit it, not merely for the production of our needs, but for the accumulation of wealth and power for the few. In the process, it has disposed of the many and has opted to replace us with cheaper and more efficient alternatives. But our newer and smarter environments in digital culture only testify to the fact that the knowledge embodied in our gadgets have now come to be totally integrated into nature itself. Which is to say that at this point in our history, no part of nature escapes the human footprint. But it shows, precisely, just how much the human impact can have upon the world. This is why Andreas Malm, for instance, proposes that "*any* call for a more environmentally beneficial practice by necessity puts humans front and centre."[25] He adds that human beings need to take responsibility for exploiting nature, not by subtracting ourselves, but by still acting upon the world. Human beings, he writes, "are the only ones who could *possibly* rise up and shake off fossil fuels from their economies. It seems a rather dispiriting and demobilising move to tell them that they are nothing special, that nothing separates them from animal or machine."[26]

Psychoanalysis and historical materialism show, not only the way that our knowledge impacts reality; both also teach us that an inherent ethics is attached to this knowledge, one that comes from our relationship with the sublime. We may act ethically only insofar as we know. The Anthropocene is now the term commonly used to prove the human impact on the planet—to show that human activities have launched the planet into a wholly new geological period. This is not all bad news. As Jameson tells us, "The glory of the Anthropocene … has been to show us that we really can change the world. Now it would be intelligent to terraform it."[27] Perhaps this is how we might imagine what Mann and Wainwright call Climate X.

Notes

1. Geoff Mann and Joel Wainwright, *Climate Leviathan: A Political Theory of Our Planetary Future* (New York: Verso, 2018).
2. Peter Frase, *Four Futures: Life After Capitalism* (New York: Verso, 2016).
3. Sigmund Freud, *Civilization and Its Discontents*, ed. Todd Dufresne, trans. Gregory C. Richter (Peterborough, ON: Broadview Press, 2016), 57.
4. Gregor Moder, *Hegel and Spinoza: Substance and Negativity* (Evanston, IL: Northwestern University Press, 2017), 76.
5. This logic is best rendered by Alenka Zupančič, *What is Sex?* (Cambridge, MA: MIT Press, 2017), 79.
6. Fredric Jameson, *The Seeds of Time* (New York: Columbia University Press, 1994), xii.
7. See, for instance, Antonio Negri, "Constituent Republic," in *Radical Thought in Italy: A Potential Politics*, ed. Paolo Virno and Michael Hardt (Minneapolis: University of Minnesota Press, 1996), 213–22; Paolo Virno, "Notes on the General Intellect," in *Marxism Beyond Marxism*,

ed. Saree Makdisi et al. (New York: Routledge, 1996), 265–72; and, Maurizio Lazzarato, "General Intellect: Towards an Inquiry into Immaterial Labour," in *Immaterial Labour, Mass Intellectuality, New Constitution, Post-Fordism and All That* (London: Red Notes, 1994).

8. Slavoj Žižek, *The Sublime Object of Ideology* (New York: Verso, 1989), 205–6.
9. Slavoj Žižek, *Sex and the Failed Absolute* (New York: Bloomsbury, 2020), 25.
10. Jacques Lacan, *The Seminar of Jacques Lacan, Book XVII: The Other Side of Psychoanalysis*, trans. Russell Grigg (New York: W. W. Norton, 2007), 35.
11. Jacques Lacan, *The Four Fundamental Concepts of Psycho-Analysis*, trans. Alan Sheridan (New York: W. W. Norton, 1977).
12. Slavoj Žižek, *Tarrying with the Negative: Kant, Hegel, and the Critique of Ideology* (Durham, NC: Duke University Press, 1993), 58–61.
13. See Todd McGowan, *Capitalism and Desire: The Psychic Cost of Free Markets* (New York: Columbia University Press, 2016).
14. See McGowan, *Emancipation After Hegel: Achieving a Contradictory Revolution* (New York: Columbia University Press, 2019).
15. Steven Shaviro, *The Universe of Things: On Speculative Realism* (Minneapolis: University of Minnesota Press, 2014), 61.
16. Slavoj Žižek, *Incontinence of the Void: Economico-Philosophical Spandrels* (Cambridge, MA: MIT Press, 2017), 228.
17. Lacan, *Book XVII: The Other Side of Psychoanalysis*, 162.
18. Žižek, *Incontinence*, 144.
19. Clint Burnham, *Does the Internet Have an Unconscious? Slavoj Žižek and Digital Culture* (New York: Bloomsbury, 2017), 52.
20. Karl Marx, *Grundrisse: Foundations of the Critique of Political Economy (Rough Draft)*, trans. Margin Nicolaus (New York: Penguin, 1993), 692.
21. Marx, *Grundrisse*, 694.
22. Nick Srnicek and Alex Williams, *Inventing the Future: Postcapitalism and a World Without Work* (New York: Verso, 2015).
23. Fredric Jameson, *Representing Capital: A Reading of Volume One* (New York: Verso, 2011).
24. Fredric Jameson, "Postmodernism, Or The Cultural Logic of Late Capitalism," *New Left Review* 1, no. 146 (1984): 92, https://newleftreview.org/issues/I146/articles/fredric-jameson-postmodernism-or-the-cultural-logic-of-late-capitalism.

25. Andreas Malm, *The Progress of This Storm: Nature and Society in a Warming World* (New York: Verso, 2018), 116; emphasis in the original.
26. Malm, *Progress of This Storm*, 118; emphasis in the original.
27. Fredric Jameson, *Allegory and Ideology* (New York: Verso, 2019), 348.

Bibliography

Burnham, Clint. 2017. *Does the Internet Have an Unconscious? Slavoj Žižek and Digital Culture*. New York: Bloomsbury.

Frase, Peter. 2016. *Four Futures: Life After Capitalism*. New York: Verso.

Freud, Sigmund. 2016. *Civilization and Its Discontents*. Edited by Todd Dufresne. Translated by Gregory C. Richter. Peterborough, ON: Broadview Press.

Jameson, Fredric. 1984. Postmodernism, Or The Cultural Logic of Late Capitalism. *New Left Review* 1 (146): 53–92. https://newleftreview.org/issues/I146/articles/fredric-jameson-postmodernism-or-the-cultural-logic-of-late-capitalism.

———. 1994. *The Seeds of Time*. New York: Columbia University Press.

———. 2011. *Representing Capital: A Reading of Volume One*. New York: Verso.

———. 2019. *Allegory and Ideology*. New York: Verso.

Lacan, Jacques. 1977. *The Four Fundamental Concepts of Psycho-Analysis*. Translated by Alan Sheridan. New York: W. W. Norton.

———. 2007. *The Seminar of Jacques Lacan, Book XVII: The Other Side of Psychoanalysis*. Translated by Russell Grigg. New York: W. W. Norton.

Lazzarato, Maurizio. 1994. General Intellect: Towards an Inquiry into Immaterial Labour. In *Immaterial Labour, Mass Intellectuality, New Constitution, Post-Fordism and All That*. Translated by Ed Emery. London: Red Notes.

Malm, Andreas. 2018. *The Progress of This Storm: Nature and Society in a Warming World*. New York: Verso.

Mann, Geoff, and Joel Wainwright. 2018. *Climate Leviathan: A Political Theory of Our Planetary Future*. New York: Verso.

Marx, Karl. 1993. *Grundrisse: Foundations of the Critique of Political Economy (Rough Draft)*. Translated by Margin Nicolaus. New York: Penguin.

McGowan, Todd. 2016. *Capitalism and Desire: The Psychic Cost of Free Markets*. New York: Columbia University Press.

———. 2019. *Emancipation After Hegel: Achieving a Contradictory Revolution*. New York: Columbia University Press.

Moder, Gregor. 2017. *Hegel and Spinoza: Substance and Negativity*. Evanston, IL: Northwestern University Press.

Negri, Antonio. 1996. Constituent Republic. In *Radical Thought in Italy: A Potential Politics*, ed. Paolo Virno and Michael Hardt, 213–222. Minneapolis: University of Minnesota Press.

Shaviro, Steven. 2014. *The Universe of Things: On Speculative Realism*. Minneapolis: University of Minnesota Press.

Srnicek, Nick, and Alex Williams. 2015. *Inventing the Future: Postcapitalism and a World Without Work*. New York: Verso.

Virno, Paolo. 1996. Notes on the 'General Intellect.'. In *Marxism Beyond Marxism*, ed. Saree Makdisi, Cesare Casarino, and Rebecca E. Karl, 265–272. New York: Routledge.

Žižek, Slavoj. 1989. *The Sublime Object of Ideology*. New York: Verso.

———. 1993. *Tarrying with the Negative: Kant, Hegel, and the Critique of Ideology*. Durham, NC: Duke University Press.

———. 2017. *Incontinence of the Void: Economico-Philosophical Spandrels*. Cambridge, MA: MIT Press.

———. 2020. *Sex and the Failed Absolute*. New York: Bloomsbury.

Zupančič, Alenka. 2017. *What is Sex?* Cambridge, MA: MIT Press.

14

From Capitalocene to Anthropocene: The Feminine Counter-Ecology of Snowpocalypse Films

Tamás Nagypál

As Fredric Jameson famously noted, we live in an era when it is easier to imagine the end of the world than the end of capitalism.[1] Post-apocalyptic films like *28 Days Later* (dir. Danny Boyle), *Snowpiercer* (dir. Bong Joon-ho), or *Mad Max: Fury Road* (dir. George Miller) are cases in point as they are set in the aftermath of a man-made ecological disaster that led to the collapse of human civilisation, leaving the survivors to battle over the remaining resources in a struggle of life and death.[2] The inhabitants of these desolate environments exist in an advanced state of *immunity*, suspending their obligation to the global commons: the social relations and modes of cooperation with each other and non-human forms of life within Earth's ecosystem—norms of coexistence that are not subordinated to the logic of capitalist exploitation.[3] In Marxist terms, these are stories of primitive accumulation: the privatisation of the commons by new barbarians through biopolitical violence, subsuming all life under the rule of a sovereign power to enforce its capitalist use or abandon it to death. To put it differently, these post-apocalyptic films explode the

T. Nagypál (✉)
Ryerson University, Toronto, ON, Canada

C. Burnham, P. Kingsbury (eds.), *Lacan and the Environment*, The Palgrave Lacan Series, https://doi.org/10.1007/978-3-030-67205-8_14

contradictions of the Anthropocene discourse, the idea that the ecological footprint of the human species inaugurated a new geological epoch on the planet. In *Inhospitable World*, Jennifer Fay points out that, on the one hand, the Anthropocene refers to humanity's modern endeavour to construct anthropogenic environments through "complete self-conditioning and liberation from nature."[4] On the other hand, considered in its totality, the result of our species' collective self-fashioning is the exact opposite: a planetary environmental catastrophe. The moment when humanity itself becomes a natural force on a global scale therefore coincides with its post-human loss of rational agency and control over the world—a paradox captured, as Phillip McReynolds notes, in the today's ubiquitous popular cultural figure of the zombie.[5] The proper term for such reduction of all human beings along with the entire ecosystem to exploitable resources is what Jason Moore calls the Capitalocene, "capitalism as a way of organizing nature—as a multispecies, situated, capitalist world-ecology,"[6] and it is this Capitalocene pushed to its world-destroying conclusion that post-apocalyptic films present as the truth of our so-called Anthropocene era.

In his *Fossil Capital: The Rise of Steam-Power and the Roots of Global Warming*, Andreas Malm builds on Moore's ideas of the Capitalocene to take issue with the assumption of the Anthropocene discourse that humanity a whole was responsible for the adaptation of fossil fuels that led to anthropogenic climate change. He emphasises that it was a small class of capitalists who decided to adapt coal powered steam engines in the late eighteenth century over the (at the time) more efficient and abundant clean energy of watermills because the new technology allowed for more spatial mobility that could in time be wagered against the political power of the working class.[7] It was therefore not necessarily the immediate interest of individual capitalists that was served by the transition, rather the long term expansion of capital itself, or as Karl Marx put it in *Capital*, capital as an "automatic subject … constantly changing from one form into the other, without becoming lost in this movement."[8] This chapter argues that, viewed through Jacques Lacan's formulas of sexuation, this post-human agency of capital, set in motion by individual capitalists but always exceeding their limitations, is a *masculine* one; it occupies the real-impossible subject position of the phallic exception that totalises the social-symbolic order of the Capitalocene by always remaining a transcendental surplus to the sum of its elements. I suggest that as long as capital occupies such a masculine topology of the constitutive exception in our imagination of the ecosystem, it is impossible to think its end.

As an alternative, I focus on a unique subgenre of snowpocalypse films that challenge the all-encompassing imaginary of the Capitalocene. These films are set in the aftermath of a—counterintuitive—total freezing of the planet, which effectively turns the ecosystem useless for capitalist exploitation, as well as for currently existing forms of human habitation. Their aesthetic of vast, horizonless white landscapes of ice and snow that dwarf the remnants of civilisation into stains on the cinematic image offer a different topology of the world, one that resists being mapped and instrumentalised from an exceptional point of view of capital's sovereign-panoptic gaze. They operate with what Fay sees as the otherworldly visuals of early documentaries about the Antarctica—a continent whose horror of "nature doomed by nature" confused its explorers' Euro-centric, universalist notions of reason, colonialism, and historical progress.[9] In snowpocalypse films, the reflexes of capitalism survive in spaces of exception, but they lose their totalising power over the Capitalocene and withdraw into collapsing shelters set against the entire Earth turned into Antarctica. This is why these films can imagine a topological alternative to our Jamesonian predicament of endlessly expanding capitalism: a counter-ecology of what Lacan called the feminine non-all where, I propose, we can find the roots of a post-capitalist Anthropocene proper. Here the subject is not placed into a state of sovereign exception with regards to the symbolic order but, in Bracha Ettinger's terms, links the porous borders of her existence with others in a shared "matrixial space" of co-belonging and co-becoming for which the mother's womb is the original model.[10]

The Frozen Capitalocene in *Quintet*

Shot in 1979 at the abandoned site of the 1967 World Expo in Montreal, Robert Altman's *Quintet*, a post-apocalyptic sci-fi noir set in a future ice age, is a powerful counterstatement to the utopian modernism of the exhibition themed Man and His World.[11] While the futuristic megastructures of Expo 67 showcased artificially engineered, self-contained habitats of modern life, in the universe of *Quintet* they become decaying ruins of what the director describes as "a dead world, in which there [is] no hope, absolutely no hope of survival."[12] An early example of postmodern environmental dystopias, the film can be productively read as a formalist exercise exposing the difficulty of imagining the apocalyptic conclusion to the Capitalocene in its totality. According to the plot, while the planet has frozen around them for unexplained reasons, the remaining survivors of humanity have withdrawn under the crumbling dome of a once thriving metropolis (set in Expo 67's Man the Explorer Pavilion) where they pass time by playing the competitive board game Quintet with regular tournaments in which players also plot to kill each other in real life. Paradoxically, the deadly game is the only widely available source of income allowing them to purchase the scarce resources they need to survive.

The dome allegorically evokes what Peter Sloterdijk calls the world interior of capital: the immunological hothouse protecting the privileged of today's global population, closing them off from the biopolitical and environmental devastation of capitalism by constructing protected micro worlds against an increasingly inhospitable outside.[13] Yet, contrary to gated communities or gentrified urban centres of our time, the film's spherical city is an autoimmune environment:[14] rather than shielding the pampered from the risk of death, it elevates the death drive into the highest principle of life. As Grigor, the game master puts it explaining the rationale of Quintet: "Life can only be felt when death is near. Every time you cheat death, you feel the pure thrill of life." The real purpose of the game is not the material reward of the winner but the possibility of playing again and again. In Lacanian terms, what therefore distracts the characters from the true weight of the catastrophe happening around them are not, as in Sloterdijk, the hedonistic pleasures of the postmodern consumer, but the *jouissance* of a death drive beyond the pleasure principle—a *jouissance* that thwarts the totalising imaginary of the snowpocalpse by both resigning to its inevitability *and* forever postponing its arrival. If this is an example of fetishistic disavowal, what is disavowed is not the negativity of death itself, but rather extinction, or, to put it differently, private death is substituted for collective death. This is the function of the domed city of Quintet players in the film: it works as an apparatus transforming the prospect of universal death into individual death, revealing the fundamental fantasy of capitalist biopolitics according to which, while individual players might lose, somebody (capital itself) always stands as the winner in the life and death competition. The ideological role of this fantasy is to prevent them as well as the viewer from facing what Ray Brassier called in his *Nihil Unbound*, the "truth of extinction,"[15] the inevitable end of humanity and indeed of Earth itself in solar death, which would make the enjoyment of death as the death drive that keeps capitalism going impossible. This hegemonic form of capitalist *jouissance*, however, is also constantly destabilised in *Quintet* by the antagonism of sexual difference.

As Joan Copjec reminds us in her *Read My Desire*, for Lacanian psychoanalysis, sex is "the stumbling block of sense," "produced by the internal limit, the failure of signification"—that coincides with the subject herself/himself.[16] This failure has to do with the symbolic order as a whole, the big Other as the set of signifiers constituting a language that is always necessarily incomplete, lacking, insofar as it is unable to signify itself (always needing one signifier more ad infinitum). This is why Lacan states that the big Other does not exist;[17] it cannot be accounted for in its totality ("there's no such thing as a metalanguage").[18] While subjects try to fill the constitutive lack of the Other they stand for with various libidinally charged fantasies, these attempts always misfire, leaving the *real* of sex to haunt the symbolic order. In his later work in introducing the formulas of sexuation (see Fig. 14.1), Lacan further elaborates on the destabilising role of sex by suggesting that the totalisation of language he refers to as the phallic function can fail in one of two incompatible ways, following either a masculine or a feminine logic that give two different topological forms to *jouissance*.[19]

As Copjec observes, the two lines of each side contradict one another, offering two antinomies that cannot be reconciled into a coherent whole neither with each other, nor with the formula aligned with them on the other side. The masculine antinomy (left) can be rendered as "There is at least one x that is not submitted to the phallic function," while at the same time "All x's are (every x is) submitted to the phallic function." By

$\exists x \; \overline{\Phi x}$	$\overline{\exists x} \; \overline{\Phi x}$
$\forall x \; \Phi x$	$\overline{\forall x} \; \Phi x$

Fig. 14.1 Formula of sexuation. (Source: author)

contrast, the feminine antinomy (right) translates to "There is not one x that is not submitted to the phallic function," yet "Not all (not every) x is submitted to the phallic function."[20]

The antinomy on the masculine side, Slavoj Žižek's *Metastases of Enjoyment* argues, applies a topology to the world that is split into two separate domains: the symbolic order and the real of *jouissance* as its constitutive exception.[21] The unique entity that is exempted from phallic accounting here is the Man of full enjoyment, like the Primal Father of the Freudian myth, whose existence marked by the real-impossible final signifier to complete the symbolic is presupposed while encounter with it is forever postponed.[22] As Copjec stresses, the masculine subject's *jouissance* does not come from embodying but from prohibiting this structurally unreachable ideal;[23] from the very imposition of a limit between the symbolic and the real that Lacan calls symbolic castration. As a result, on one level, the masculine universe seeks to include all phenomena while at another it needs to remain fundamentally incomplete.

The universe of *Quintet* exemplifies the masculine topology of the symbolic pushed to a point of autoimmune breakdown. Here human beings are accounted for by the phallic function, or in other words, come to *exist* in the symbolic order of the dying ecosystem at the moment of their death, and those still living are placed in a temporary state of exception from this norm. By playing the game, however, they are also identified, that is to say, killed by the phallic apparatus of Quintet one by one—all except for the winner of every cycle ad infinitum. At the beginning of the film, we see the protagonist Essex and his wife Vivia pass by several frozen human corpses being eaten by dogs as they are making their way to the city—a site of exception from the universe of death. Once inside, however, dead bodies keep popping up around them—most of them former players in the game—suggesting a destabilising boundary between the inside and the outside, or between the symbolic and the real. For instance, when they find shelter at the home of Essex's brother, Vivia stays with the four-member family to play Quintet while her husband goes out to buy firewood. As they are playing, someone opens the door behind them and throws a bomb into the room, killing everyone from yet another space of phallic exception, only to have his throat slit a few minutes later by another player sneaking up on him from a dark alley. The game of Quintet itself is also an exercise in the masculine topology of exception. As the name suggests, it is a five-player game but it has an additional role for a "sixth man" who has the power to impose his will on the players from the outside by "arranging the killing order." The ultimate sixth man of the film is Grigor, the master of Quintet tournaments who never plays himself but who has the function of interpreting the rules, which in practice translates to him encouraging players to murder each other when they have the upper hand. As he puts it, his job is to keep the game alive—an allegory for the automatic subjectivity of capital's death drive.

Crucially, we never learn the actual rules of Quintet—a fact that disappointed critics like Roger Ebert who was frustrated by the film's incoherence.[24] From a Lacanian perspective, however, this inexplicability of the game points at the antagonism of sexual difference at the core of the film's totalising imaginary. If masculine *jouissance* is identified with imposing death, the exhaustion of someone's resource value, from the position of the sixth man forever exempted from it, the possibility of another, feminine enjoyment emerges through the form of collective play that escapes this topology of phallic separation. As Copjec emphasises, it is not that the feminine logic can somehow complete the symbolic order, only that the final, primordially repressed signifier covering its constitutive lack—the perennial sixth man of Quintet—is not assumed to exist anymore, which is why the game—and by extension the universe of the film—then appears *inconsistent*, non-all, rather than *incomplete*.[25] In other words, femininity is not the real-impossible being posited outside the symbolic but one fully immanent to it: it is the name for the symbolic order's inherent deadlock.[26] The feminine topology therefore maps the "indeterminate existence" of those belonging to an always open, "infinite set" instead of the finite one closed off by an exceptional signifier.[27]

The film's two female protagonists express a desire for this feminine existence but they are not able to break away from the masculine universe. Vivia, an example of the noir genre's nurturing woman stereotype,[28] is marked by the phallic exception as she is pregnant with the only child on the planet. Her sexual difference proper from her husband is emphasised, however, in the scene when Essex is trying to locate his brother in the city's directory (with the help of five numbers and an additional colour code), while she decides to play hide and seek in the room full of fractured glass plates evoking the classical noir finale of *The Lady from Shanghai* in a hall of mirrors. As Copjec stresses, contrary to her masculine counterpart, the feminine subject does not properly exist in the symbolic as she cannot be re-found, because she occupies an inconsistent space in which a judgment of existence is impossible.[29] Vivia, however, is made to "exist" when she enters the phallic topology of Quintet and is killed as a result. The film's other female noir archetype, Ambrosia, is a reluctant *femme fatale* who likes to play but does not enjoy the role of the phallic exception: "being a sixth man is an advantage but it's also boring," she says, "I just start to get excited and it's all over." By contrast, Altman differentiates the intimate enjoyment of her sleeping with Essex through haptic imagery, with the camera gradually moving from an overhead shot of the two crawled under a blanket in the ice cold room to extreme close-ups of parts of their faces, hands, and skin as they kiss and caress each other. Ettinger calls this the feminine *jouissance* of "matrixial borderlinking" where "the private subjectivity of the individual is momentarily unbounded. The psyche momentarily melts, and its psychic threads are interwoven with threads emanating from objects, images, and other subjects."[30] As Ettinger stresses, such encounter is always temporary and partial, and should not be confused with psychosis where the phallic function of symbolic castration fails altogether and the subject cannot establish any boundaries between herself and the Other.[31] Accordingly, Essex and Ambrosia are back to playing Quintet the next day; she attacks him, only to be killed by the man's motoric gesture of self defense in the final round of the tournament.

It is therefore Essex, an impostor masquerading as the Quintet player who killed his wife and brother, who wins the game he only joined to solve a murder mystery. As often in film noir, the answer to his investigation turns out to be self-referential: it is he himself, or more precisely the place he occupies in the masculine topology, that is responsible for the killings. When he learns this, with a "courage of hopelessness,"[32] he refuses to continue playing and decides to leave the city and go north, despite Grigor's insistence that there is nothing there but death. He doesn't die, however, rather disappears into the vast frozen landscape, and the film ends with a white blur with no clearly discernible figures in the frame. The camera lens itself, blurred around the edges due to a layer of Vaseline throughout the film, now seems to freeze over entirely, forming a haptic surface for the viewer to "touch" with her eyes rather than an optical experience of panoptic control.[33] Essex's existence, along with everything else in the diegetic universe, becomes indeterminate, signalling that we had crossed over to a feminine topology of the symbolic where the viewer can finally face the truth of extinction, the inevitable end of the Capitalocene, freeing her imagination to think its alternative.

The Day after the Capitalocene in *The Wandering Earth*

Many contemporary climate fiction films feature the ecological disaster of a snowpocalypse. Some, like *The Day After Tomorrow*, directed by Roland Emmerich or *Sunshine*, directed by Danny Boyle, evoke its possibility as a cautionary tale to urge global action and prevent the destruction of human civilisation in its current state, while others like *The Colony*, directed by Jeff Renfro and Bong Joon-ho's *Snowpiercer*, follow the post-apocalyptic formula of *Quintet* in embracing the radical negativity of the ice age as an exodus from capitalism.[34] The recent Chinese blockbuster *The Wandering Earth*, directed by Frant Gwo, goes a step further in imagining life on Earth after capitalism is already over and humanity is facing the challenge of the Anthropocene proper: how to organise itself into a collective force capable of shaping the ecosystem into a habitat that would protect everyone—a project Sloterdijk calls co-immunism.[35] Crucially, the crisis they face is not simply climate change but extinction brought about by the death of the Sun. At the beginning of the film, half of the world's population along with its natural ecosystem is already dead, and the survivors have set aside their conflicts to build hundreds of "Earth engines" that would propel the planet into another solar system. As a result of leaving its orbit around the Sun, the planet has frozen over and humanity has withdrawn into underground mega-cities with artificial atmospheres.

While capitalism is officially gone, its topology haunts the film through immunised spaces of exception. The central conflict of the narrative emerges when Earth, on its way out of the solar system, gets too close to Jupiter's field of gravity that threatens to suck its entire mass into itself, putting an end to the Wandering Earth project—a literalisation of the popular ecological metaphor "Spaceship Earth" referring to humanity's collective responsibility for the ecosystem.[36] The dilemma is whether to have the "International Navigation Platform Space Station" abandon the planet and fly away with a fraction of humanity in a hibernated state, or use the station's navigation panel to crash into Jupiter and ignite an explosion whose shockwaves could push Earth back into a safe trajectory, but could also fail and therefore jeopardise the survival of the human race. In Lacanian terms, this setup resembles the *vel* of castration, the forced choice between *jouissance* and social-symbolic existence where full enjoyment is always already lost: the choice of uncastrated *jouissance* (here: saving the whole of humanity) is not a real one as it leads to psychosis, the total erasure of social-symbolic identity (extinction).[37] In the film, however, this binary logic is merely the limited perspective of the "villain," the AI running the space station. After assessing the situation from its panoptic position,[38] it makes a sovereign decision to sacrifice the planet, which turns its totalising masculine gaze into a world destroying one, akin to the automatic subjectivity of capital. Meanwhile, on the surface of the planet we see the same topological tension with the young male protagonist Liu Qi taking his grandfather's truck—used to carry fuel to the Earth engines—for a joyride with his sister Han Duoduo. They enjoy their illusion of safety in the micro-atmosphere of the vehicle, carelessly bumping into various machines that work to keep the planet flying. In a thinly veiled didactic moral twist condemning bourgeois individualism in favour of Chinese collectivism, the kids are soon arrested, and find themselves entangled in a global operation with thousands of trucks like theirs carrying a load to restart the engines of Earth.

The film is consistent in problematising private atmospheres of exception, whether it's the space station with its hibernation pods, the truck of runaway children, or the air-conditioned suit and helmet individuals have to wear in the frozen and oxygen deprived environment of Earth's surface. As the planet becomes increasingly unstable due to Jupiter's pull on its atmosphere, the boundaries of people's private life-bubbles also become more porous: windshields of vehicles break, the glass helmets crack, oxygen tanks run out, and the protective shielding around underground cities crumbles. While the official narrative is about the restoration of global atmospheric control, at a libidinal level the film offers the viewer the feminine *jouissance* of "matrixial borderlinking"[39] via the experience of shared vulnerability and co-immunity as the necessary precondition of such a collective effort. This is summed up in the scene when Liu Qi and Han Duoduo, while heroically assisting an international team in restarting one of the Earth engines to propel the planet away from Jupiter, fall from a collapsing bridge while their helmets break from falling debris. A slow-motion sequence shows that they manage to hold onto each other and activate a womb-like inflatable sphere saving them from being crushed.

Conclusion

Snowpocalypse films demonstrate that imagining the end of the Capitalocene is a topological problem. Viewed through the masculine lens of the constitutive exception, the end of the world is the real-impossible event that is always approaching but never arrives; it is the moment when capital's immunity from responsibility for the ecosystem is devastating and would reach an auto-immune state of collapse with the extinction of humanity. Symptomatically, agents of the phallic exception like *Quintet*'s cynical game master and *The Wandering Earth*'s genocidal AI seek to postpone the apocalypse through the very acts of destruction that brings it closer. The privatisation of immunity, the construction of atmospheres of exception against the ecosystem as a whole is their method of survival. By contrast, shifting to a feminine topological perspective in these films allows for passing through the apocalyptic zero point and invalidate its pull by taking the environment to be both already "dead," that is, useless for capitalist exploitation, and inescapable. Contrary to the masculine forced choice between the endless privatisation of the atmosphere or death, the feminine alternative is the co-immunism of the Anthropocene proper. It reimagines immunity not as the suspension of one's obligation to a collective, but as a matter of always already shared co-belonging and collective vulnerability in overlapping and intermingling atmospheric bubbles that cannot be privately controlled by any sovereign power.[40] Beyond their narrative trajectories critiquing apparatuses of exception, the aesthetic of snowpocalypse films also highlights co-immunism in a variety of ways. *Quintet* mobilises increasingly abstract images of frozen landscapes to facilitate the viewer's haptic encounter with the surface of the screen as the interface with the diegetic world. *The Wandering Earth* offers CGI spectacles of various fragile atmospheres interacting with one another, perhaps most notably Earth "borderlinking" with Jupiter. Similarly, the final shot of the film shows Liu Qi and Han Duoduo, now responsible adults, in the driver's cabin of truck just like the one they stole when they were younger. As they start their daily routine of maintaining the Earth Engines, the virtual camera tracks back without a cut, travelling through the frame of their vehicle to show a vast network of service routes with hundreds of trucks covering the surface of the frozen planet. We then move further up, following the beams of the engines through layers of clouds to settle on the image of the Wandering Earth flying through the cosmos, leaving a vapour trail behind. Meanwhile, the voiceover narration speaks of the 2100 year voyage of "hope through despair" mankind embarked upon: a fitting description of the Anthropocene of co-immunism.

Notes

1. See Fredric Jameson, "The Antinomies of Postmodernity," in *The Cultural Turn: Selected Writings on the Postmodern, 1983-1998*, by Fredric Jameson (New York: Verso, 1998), 50.
2. See Films: Alex Garland, *28 Days Later*, directed by Danny Boyle (London: DNA Films, 2002); Bong Joon-ho and Kelly Masterson, *Snowpiercer*, directed by Bong Joon-ho (Seoul, South Korea: Moho Films, 2013); and George Miller, Brendan McCarthy, and Nico Lathouris, *Mad Max: Fury Road*, directed by George Miller (Burbank, CA: Warner Bros., 2015).
3. For a concept of immunity as the suspension of one's obligation to the commons, see Roberto Esposito, *Bios: Biopolitics and Philosophy*, trans. Timothy Campbell (Minneapolis: University of Minnesota Press, 2008), 45-78.
4. Jennifer Fay, *Inhospitable World: Cinema in the Time of the Anthropocene* (Oxford: Oxford University Press, 2018), 14.
5. Phillip McReynolds, "Zombie Cinema and the Anthropocene: Posthuman Agency and Embodiment at the End of the World," *Cinema: Journal of Philosophy and the Moving Image* 7 (2015): 149-68, http://cjpmi.ifilnova.pt/storage/7/7_McReynolds.pdf (accessed June 7, 2020).
6. Jason W. Moore, "Anthropocene or Capitalocene? Nature, History, and the Crisis of Capitalism," in *Anthropocene or Capitalocene? Nature, History, and the Crisis of Capitalism,* ed. Jason W. Moore (Oakland, CA: PM Press, 2016), 6.
7. Andreas Malm, *Fossil Capital: The Rise of Steam Power and the Roots of Global Warming* (London and New York: Verso, 2016).
8. Karl Marx, *Capital: A Critique of Political Economy, Volume 1.*, trans. Ben Fowkes (London: Penguin Books, 1976), 255.
9. Fay, *Inhospitable World*, 182.
10. Bracha L. Ettinger, *The Matrixial Borderspace*, ed. Brian Massumi (Minneapolis: University of Minnesota Press, 2006).
11. Robert Altman, Frank Barhydt, and Patricia Resnick, *Quintet*, directed by Robert Altman (Santa Monica, CA: Lion's Gate, 1979).
12. Robert Altman, quoted in Steve Lyons, "Memory of a Post-Apocalyptic Future: Whitening Skeletons and Frozen Time in Robert Altman's *Quintet* and Expo 67's Man the Explorer Pavilion," paper presented at the Montreal as Palimpsest II: Hauntings, Occupations, Theatres of Memory conference, Department of Art History, Concordia University,

Montreal QC, April 17, 2009, https://cityaspalimpsest.concordia.ca/palimpsest_II_fr/papers/Steve_Lyons.pdf (accessed June 7, 2020).
13. See Peter Sloterdijk, *In the World Interior of Capital: For a Philosophical Theory of Globalization*, trans. Wieland Hoban (Cambridge: Polity Press, 2013).
14. Esposito defines autoimmunity as the escalation of the tendencies already present in immunity: a "condition in which the protective apparatus becomes so aggressive that it turns against its own body (which is what it should protect), leading to its death" (Esposito, *Bios*, 116).
15. See Ray Brassier, *Nihil Unbound: Enlightenment and Extinction* (New York: Palgrave Macmillan, 2007), 205-68.
16. Joan Copjec, *Read My Desire: Lacan against the Historicists* (Cambridge, MA: MIT Press, 1994), 204.
17. Jacques Lacan, "The Subversion of the Subject and the Dialectic of Desire in the Freudian Unconscious," in *Écrits: The First Complete English Edition*, trans. Bruce Fink (New York: W. W. Norton, 2006.), 700.
18. Jacques Lacan, *The Seminar of Jacques Lacan, Book XX: On Feminine Sexuality: The Limits of Love and Knowledge, Encore, 1972-1973*, ed. Jacques-Alain Miller, trans. Bruce Fink (New York, W. W. Norton, 1998), 107.
19. Lacan, *Book XX: On Feminine Sexuality*, 56-57.
20. Copjec, *Read My Desire*, 214.
21. Slavoj Žižek, *The Metastases of Enjoyment: Six Essays on Women and Causality* (New York: Verso, 1994), 153.
22. In his *Seminar XIX*, Lacan gives the example of the Freudian father of the primal horde who enjoys all the women as the paradigmatic case of phallic exception. See Jacques Lacan, *The Seminar of Jacques Lacan, Book XIX: ...or Worse*, ed. Jacques-Alain Miller, trans. A. R. Price (Cambridge: Polity, 2018), 78-95.
23. Copjec, *Read My Desire*, 235.
24. Roger Ebert, "Quintet." *RogerEbert.com*, February 16, 1979, https://www.rogerebert.com/reviews/quintet (accessed October 1, 2019).
25. Copjec, *Read My Desire*, 231.
26. As Lacan puts it, femininity is a "being [that] has no other locus than the locus of the Other" (Lacan, *Book XX: On Feminine Sexuality*, 77).
27. Lacan, *Book XX: On Feminine Sexuality*, 103.
28. On the noir stereotypes of the "nurturing woman" and its opposite, the "spider woman," see Janey Place, "Women in Film Noir," in *Women in Film Noir*, rev. ed., ed. E. Ann Kaplan (London: British Film Institute, 1998), 47-69.

29. Copjec, *Read My Desire*, 221.
30. Bracha L. Ettinger, "*Fascinance* and the Girl-to-m/Other Matrixial Feminine Difference," in *Psychoanalysis and the Image: Transdisciplinary Perspectives*, ed. Griselda Pollock (Malden, MA: Blackwell Publishing, 2006), 62.
31. Ettinger, "*Fascinance*," 84.
32. See Slavoj Žižek, *The Courage of Hopelessness: Chronicles of a Year of Acting Dangerously* (London: Penguin, 2017); Žižek is not advocating apocalyptic nihilism here, but argues that one has to pass through a zero point of hopelessness, a radical libidinal divestment from the project of amending the current global capitalist status quo in order to imagine its truly revolutionary alternative. Similarly, *Quintet* leaves open the possibility of a new world after the destruction of the present one in an early image of a goose flying north—a move that seems irrational at the time to Essex who is still headed the opposite direction to the shelter of the domed city.
33. On such feminine aesthetics of haptic visuality, see Laura U. Marks, "Video Haptics and Erotics," *Screen: The Journal of the Society for Education in Film and Television* 39, no. 4 (1998): 331-48, 10.1093/screen/39.4.331.
34. See films: Roland Emmerich, Roland, and Jeffrey Nachmanoff, *The Day After Tomorrow*, directed by Roland Emmerich (Los Angeles: Twentieth Century Fox, 2004); Alex Garland, *Sunshine*, directed by Danny Boyle (London: DNA Films, 2007); Jeff Renfroe, Svet Rouskov, Patrick Tarr, and Pascal Trottier, *The Colony*, directed by Jeff Renfroe (Woodland Hills, CA: RLJ Entertainment, 2013); and Bong Joon-ho's *Snowpiercer*. *Snowpiercer* especially has been analysed by many in this context, see Gerry Canavan, "'If the Engine Ever Stops, We'd All Die': *Snowpiercer* and Necrofuturism," *Paradoxa: Studies in World Literary Genres* 26 (2014): 41-66, https://epublications.marquette.edu/cgi/viewcontent.cgi?article=1298&context=english_fac; and Fred Lee and Steven Manicastri, "Not All Are Aboard: Decolonizing Exodus in Joon-ho Bong's *Snowpiercer*," *New Political Science* 40, no. 2 (2018): 211-26, 10.1080/07393148.2018.1449405.
35. See film: Gong Ge'er, Junce Ye, Yan Dongxu, Frant Gwo, and Yang Zhixue, *The Wandering Earth*, directed by Frant Gwo (Beijing: China Film Group, 2019); and see also Peter Sloterdijk, *Bubbles: Microspherology*, trans. Wieland Hoban (Los Angeles: Semiotext(e), 2011).
36. See R. Buckminster Fuller, *Operating Manual for Spaceship Earth* (Carbondale: Southern Illinois University Press, 1969).

37. On Lacan's notion of the forced choice, see Jacques Lacan, *The Seminar of Jacques Lacan, The Seminar of Jacques Lacan, Book XI: The Four Fundamental Concepts of Psychoanalysis*, ed. Jacques-Alain Miller, trans. Alan Sheridan (New York: W. W. Norton, 1981), 210-13.
38. The perspective of the station is introduced through a rapid digital tracking shot where the virtual camera travels from the surface of the Earth thousands of kilometers into space in a few seconds, settling on a point of view from which the station dwarves the tiny blue marble of the planet, and looks as a giant weapon directed at it.
39. Ettinger, *Matrixial Borderspace.*
40. In their Lacanian topological critique of the exceptional status of surplus under capitalism, Ceren Özselçuk and Yahya M. Madra offer a similar definition of communism as the "axiom that asserts that no one can have exclusive rights over the dispatching of the surplus." See Ceren Özselçuk and Yahya M. Madra, "Psychoanalysis and Marxism: From Capitalist-All to Communist Non-All," *Psychoanalysis, Culture & Society* 10, no. 1 (2005): 93, 10.1057/palgrave.pcs.2100028.

Filmography

Altman, Robert, Frank Barhydt, and Patricia Resnick. 1979. *Quintet.* Directed by Robert Altman. Santa Monica, CA: Lion's Gate.

Bong, Joon-ho, and Kelly Masterson. 2013. *Snowpiercer.* Directed by Bong Joon-ho. Seoul, South Korea: Moho Films.

Emmerich, Roland, and Jeffrey Nachmanoff. 2004. *The Day After Tomorrow.* Directed by Roland Emmerich. Los Angeles: Twentieth Century Fox.

Garland, Alex. 2002. *28 Days Later.* Directed by Danny Boyle. London: DNA Films.

———. 2007. *Sunshine.* Directed by Danny Boyle. London: DNA Films.

Gong, Ge'er, Junce Ye, Yan Dongxu, Frant Gwo, and Yang Zhixue. 2019. *The Wandering Earth.* Directed by Frant Gwo. Beijing: China Film Group.

Miller, George, Brendan McCarthy, and Nico Lathouris. 2015. *Mad Max: Fury Road.* Directed by George Miller. Burbank, CA: Warner Bros.

Renfroe, Jeff, Svet Rouskov, Patrick Tarr, and Pascal Trottier. 2013. *The Colony.* Directed by Jeff Renfroe. Woodland Hills, CA: RLJ Entertainment.

Bibliography

Brassier, Ray. 2007. *Nihil Unbound: Enlightenment and Extinction*. New York: Palgrave Macmillan.

Canavan, Gerry. 2014. 'If the Engine Ever Stops, We'd All Die': *Snowpiercer* and Necrofuturism. *Paradoxa: Studies in World Literary Genres* 26: 41–66. https://epublications.marquette.edu/cgi/viewcontent.cgi?article=1298&context=english_fac.

Copjec, Joan. 1994. *Read My Desire: Lacan against the Historicists*. Cambridge, MA: MIT Press.

Ebert, Roger. 1979. "Quintet." *RogerEbert.com*. February 16. https://www.rogerebert.com/reviews/quintet. Accessed 1 Oct 2019.

Esposito, Roberto. 2008. *Bios: Biopolitics and Philosophy*. Trans. Timothy Campbell. Minneapolis: University of Minnesota Press.

Ettinger, Bracha L. 2006a. *Fascinance* and the Girl-to-m/Other Matrixial Feminine Difference. In *Psychoanalysis and the Image: Transdisciplinary Perspectives*, ed. Griselda Pollock, 60–94. Malden, MA: Blackwell.

———. 2006b. In *The Matrixial Borderspace*, ed. Brian Massumi. Minneapolis: University of Minnesota Press.

Fay, Jennifer. 2018. *Inhospitable World: Cinema in the Time of the Anthropocene*. Oxford: Oxford University Press.

Fuller, R. Buckminster. 1969. *Operating Manual for Spaceship Earth*. Carbondale: Southern Illinois University Press.

Jameson, Fredric. "The Antinomies of Postmodernity." In *The Cultural Turn: Selected Writings on the Postmodern, 1983-1998*, by Fredric Jameson, 50-72. New York: Verso, 1998.

Lacan, Jacques. 1981. *The Seminar of Jacques Lacan, Book XI: The Four Fundamental Concepts of Psychoanalysis*. Ed. Jacques-Alain Miller. Trans. Alan Sheridan. New York: W. W. Norton.

———. 1998. *The Seminar of Jacques Lacan, Book XX: On Feminine Sexuality: The Limits of Love and Knowledge, Encore, 1972-1973*. Edited by Jacques-Alain Miller. Trans. Bruce Fink. New York, W. W. Norton.

———. 2006. The Subversion of the Subject and the Dialectic of Desire in the Freudian Unconscious. In *Écrits: The First Complete English Edition*. Trans. Bruce Fink, 671-702. New York: W. W. Norton.

———. 2018. *The Seminar of Jacques Lacan, Book XIX: ...or Worse*. Edited by Jacques-Alain Miller. Trans. A. R. Price. Cambridge: Polity Press.

Lee, Fred, and Steven Manicastri. 2018. Not All Are Aboard: Decolonizing Exodus in Joon-ho Bong's *Snowpiercer*. *New Political Science* 40 (2): 211–226. https://doi.org/10.1080/07393148.2018.1449405.

Lyons, Steve. 2009. Memory of a Post-Apocalyptic Future: Whitening Skeletons and Frozen Time in Robert Altman's *Quintet* and Expo 67's Man the Explorer Pavilion. Paper presented at the Montreal as Palimpsest II: Hauntings, Occupations, Theatres of Memory conference. Department of Art History. Concordia University, Montreal, QC. April 17. https://cityaspalimpsest.concordia.ca/palimpsest_II_fr/papers/Steve_Lyons.pdf. Accessed 7 Jun 2020.

Malm, Andreas. 2016. *Fossil Capital: The Rise of Steam Power and the Roots of Global Warming*. London and New York: Verso.

Marks, Laura U. 1998. Video Haptics and Erotics. *Screen: The Journal of the Society for Education in Film and Television* 39 (4): 331–348. https://doi.org/10.1093/screen/39.4.331.

Marx, Karl. 1976. *Capital: A Critique of Political Economy, Volume 1*. Trans. Ben Fowkes. London: Penguin Books.

McReynolds, Phillip. 2015. Zombie Cinema and the Anthropocene: Posthuman Agency and Embodiment at the End of the World. *Cinema: Journal of Philosophy and the Moving Image* 7: 149–168. http://cjpmi.ifilnova.pt/storage/7/7_McReynolds.pdf. Accessed 7 Jun 2020.

Moore, Jason W. 2016. Anthropocene or Capitalocene? Nature, History, and the Crisis of Capitalism. In *Anthropocene or Capitalocene? Nature, History, and the Crisis of Capitalism*, ed. Jason W. Moore, 1–14. Oakland, CA: PM Press.

Özselçuk, Ceren, and Yahya M. Madra. 2005. Psychoanalysis and Marxism: From Capitalist-All to Communist Non-All. *Psychoanalysis, Culture & Society* 10 (1): 79–97. https://doi.org/10.1057/palgrave.pcs.2100028.

Place, Janey. 1998. Women in Film Noir. In *Women in Film Noir*, Revised ed., ed. E. Ann Kaplan, 47-69. London: British Film Institute.

Sloterdijk, Peter. 2011. *Bubbles: Microspherology*. Trans. Wieland Hoban. Los Angeles: Semiotext(e).

———. 2013. *In the World Interior of Capital: For a Philosophical Theory of Globalization*. Trans. Wieland Hoban. Cambridge: Polity Press.

Žižek, Slavoj. 1994. *The Metastases of Enjoyment: Six Essays on Women and Causality*. New York: Verso.

———. 2017. *The Courage of Hopelessness: Chronicles of a Year of Acting Dangerously*. London: Penguin.

15

Self-Destruction and the Natural World

Todd McGowan

When we think about our relationship to the world, the grandeur of the world seems to dwarf our own. We are small finite beings in a vast universe that extends far beyond our perceptive capacities. Gaining a proper understanding of this relationship would thus appear to depend on an adequate sense of our smallness in relation to the universe in which we participate. The relationship itself seems to cry out for the humility of the subject trying to understand it. It is my contention, however, that the failure to recognise the priority of the subject in relation to the universe of objects cause us to misunderstand both. It causes us to miss the traumatic cut that subjectivity marks in the natural world and that alienates speaking subjects from any animal needs. We can only grasp our relationship to the universe if we examine it through the excessiveness of subjectivity that distorts this relationship. This is what Immanuel Kant fails to do in his famous statement near the end of the *Critique of Practical Reason*.

T. McGowan (✉)
University of Vermont, Burlington, VT, USA
e-mail: todd.mcgowan@uvm.edu

C. Burnham, P. Kingsbury (eds.), *Lacan and the Environment*, The Palgrave Lacan Series, https://doi.org/10.1007/978-3-030-67205-8_15

The Lure of Starry Skies

In the conclusion of this work of moral philosophy, Kant compares our relationship to the natural world with our relationship to the moral law. He claims, "Two things fill the mind with ever new and increasing admiration and reverence, the more often and more steadily one reflects on them: *the starry heavens above me and the moral law within me*."[1] In some sense, on the one hand, our relationship to both the natural world and the moral law is the same: they fill the subject with wonder. But on the other hand, they are completely opposed, each placing the subject in a radically different relationship to what it perceives. Looking at the starry heavens reminds Kant of his smallness in the universe, while the moral law reminds him of the subject's grandeur, that subjectivity is capable of a transcendence that the natural world is not. Reconciling these two competing observations goes beyond Kant himself, who never sees the contradiction in them because he believes that subject can successfully inhabit two different worlds—a world of appearances and a supersensible world that is home to the subject's moral activity. The two observations apply to two different aspects of subjectivity. But Georg Wilhelm Friedrich Hegel and Jacques Lacan, in different ways, make possible a reconciliation of the observations without splitting appearances from the supersensible world. Only by making this move can we make sense of the relationship between subjectivity and the natural world. Kant's two experiences of wonder point in opposite directions for understanding this relationship, which is why we must decipher which one has priority.

The order that Kant frames these two experiences is important for how he thinks of subjectivity. In the first statement, Kant expresses his smallness in the face of the multitude of other worlds that he sees in the nighttime sky. The starry heavens above reveal to Kant that he is an insignificant animal that emerged out of the immense physical universe and will soon return to it without making much of an impact. He is first and foremost a finite being.[2] Kant takes his own finitude as the primary form of subjectivity, as the basis for how subjectivity intrinsically relates to the natural world. This vision of the natural world fills him with reverence because it functions perfectly without him. His awe is consummate with his own insignificance. Although contemporary critics of Kant accuse him of disavowing human finitude through his overly grandiose conception of subjectivity, this statement from the *Critique of Pure Reason* makes it clear that the finitude represents Kant's point of departure.

But then Kant has an experience that reverses the situation completely. When he reflects on the moral law within, he feels a similar reverential awe, but rather than grasping his smallness in relation to the natural world, he sees in his encounter with the moral law his elevation over all merely physical being. The location of the object that occasions that awe undergoes a dramatic reversal. Kant moves from the world outstripping the subject to the subject outstripping the world. The moral law enables the subject to act with indifference relative to all the claims that the world can make on it. Because I have the moral law, I can opt to act against societal pressures, natural impulses, and even the threat of death. The moral law enables me to act against my own good. Although it is a constraint, the moral law has the effect of freeing me from all other constraints, which is why Kant reveres it to the extent that he does.[3]

One might imagine an objection to Kant for the haughtiness of this transition from the reverence of nature to reverence before the moral law within. From being a humble creature in a vast universe in which he is just a part, he becomes a subject who towers over the entire physical universe. The arrogance of Western philosophy has never seemed so explicitly on display as it is in this brief passage. According to this reading, Kant's first insight into his insignificance is correct, but his second is a delusion of grandeur that sets German Idealism off on exactly the wrong track.[4]

But Kant's error in the conclusion of the *Critique of Practical Reason* does not occur here. It is not Kant's philosophical arrogance that leads him astray but his initial humility, his presupposition of his own finitude that inspires wonder in the vastness of the heavens. By giving a chronological priority to his existence as what he calls an "animal creature" that has a minimal ontological significance in the universe, Kant actually denigrates the effect of the subject and promulgates an imaginary vision of his own animality, a vision that fails to take into effect the distortion that signification—and the subject of the signifier—creates in the world. Finite animal being is an imaginary lure that disguises the cut of signification that denaturalises the subject and renders it excessive.[5] Kant's error here is an idiosyncratic misstep but plays a paradigmatic role in thinking about the subject's relationship to its natural being, a way of thinking that has become much more pronounced than in Kant's epoch. This imaginary conception of human animality increasingly befuddles us today. Kant is not the inventor of it, but since he grasps so clearly the excessiveness of subjectivity deriving from the moral law, more than anyone else he should have seen that human animality serves as a lure for hiding how subjectivity actually functions. My contention is that the image of the human animal disavows the excess of subjectivity that the subjection to the signifier produces and hides the trauma inherent in this excess.

What Kant should have seen is that it is the moral law within—that is, the subject's ability to act against its own interest, its drive to destroy itself—that is the condition of possibility for grasping itself as an insignificant animal creature. In other words, it is our infinitude that causes our finitude to become conspicuous for us.[6] Other animal creatures, as far as we know, have no sense of their own insignificance. They exist within their worlds in a way that the subject cannot. It is the deforming power of the moral law that renders the subject capable of considering itself in its animality. That is to say, it is what alienates the subject from its animal being that is retroactively the source of its animal being as a type of being that the subject can be aware of. As a result, one cannot separate the subject's animal being from subjectivity. It is not as if we are both animals and subjects at different points of our existence but that subjectivity colours and disfigures all animality. We cannot think of the subject's animality or the natural world as distinct from how we think of subjectivity itself.

Animality is not a neutral state prior to subjectivity but an aftereffect (an effect of what Freud calls *Nachträglichkeit*) of the emergence of the subject. The subject is not first an animal and subsequently a subject. Even prior to its birth, the subject exists in a bath of signification that denaturalises it and the world that it enters. The subject's natural being or animality is itself the posited presupposition of subjectivity. Rather than seeing animality as an initial state that subjectivity disrupts as Kant does, we must recognise the priority of subjectivity and conceive animality as subjectivity's imaginary origin. In other words, as the famous passage on the starry heavens above and the moral law within shows, Kant minimises the effect of subjectivity rather than overblowing it by disavowing the role that subjectivity has in forming our animality.

This disavowal has become much more widespread in the years since Kant wrote the *Critique of Practical Reason* and has come to completely dominate—and disfigure—our thinking about how we fit in (or do not fit in) the natural world. It is Kant's inability to see the priority of the subject relative to the animal, not his elevation of the subject above the world of objects, that leads to the great error that haunts modernity: the tendency to distinguish between nature and culture when talking about subjectivity. Ironically, it is not Kant's transcendentalism that causes the problem, but his failure to take his transcendentalism to its end point. The contemporary critique of transcendental thinking does not point towards a solution but is itself a symptom of the problem. In order to understand the subject's position relative to the world, we need more transcendentalism rather than less.

Even those who see the subject as an amalgamation of nature and culture, nonetheless err by characterising these forces as distinct and independent. Because his conception of the uprooting effect of the moral law and subjectivity is not exacting enough—because he is not enough of a transcendental thinker—Kant leaves space open for those who would analyse the subject's actions as those of a self-interested animal, which is a catastrophic error. Kant's initial image of the human animal bespeaks his failure to see the radicality of the break that the moral law introduces. It is true that one can always look back at the human animal from the position of subjectivity, but in doing so, one must take into account the fundamental distortion that subjectivity introduces into this look. The subject can only look through its own excessiveness at what it sees. What appears as a human animal is the product of a desire for this figure, not its immediate presence within the field of vision. As a result of its alienation through the signifier, subjectivity can never look directly on the world. Rather than taking an objective or neutral look, it always sees through the mediating power of signification and sees through its own desire.

When Kant discovers the role that the moral law plays for the subject, he does not yet fully comprehend the magnitude of this discovery. He does not just find that the subject can rely on reason to guide its deeds, but also that the moral law signals the existence of a principle of self-destructiveness in the subject, what Sigmund Freud and Jacques Lacan call the drive.[7] Like the moral law, the drive produces satisfaction through the sacrifice of the subject's good, through its destruction of self-interest. The moral law imposes itself on the subject's natural being and completely reshapes this being by introducing a foreign element into it, a foreign element that leads the subject to find its satisfaction in attacking its own self-interest rather than following it.[8] If one sees Kant's discovery in this light, the importance of the Second Critique becomes exponentially greater.

It becomes clear that Kant anticipates Freud's 1920 insight into the subject's unconscious self-destructiveness. The Kantian moral law is the true precursor of the Freudian death drive, not the Freudian superego, which is precisely not what one would expect. The moral law, as Kant conceives it, enables the subject to act against its own interest and the interest of others without concern for what it will ultimately gain from moral action. This last point is crucial for Kant. He brooks no possibility for the subject gaming its obedience to the moral law in order to receive an ultimately greater payoff of self-interest in the end. Any transactional morality in which the subject gains as a result of its moral action represents a complete betrayal of the moral law. Morality is important for Kant because it offers freedom from the logic of quid pro quo that determines so many of every interactions (especially in the capitalist universe).

Those who would dismiss the moral law as a version of the reality principle—I act against my self-interest now for the sake of a later reward—completely misunderstand what Kant is up to with his conception of the moral law. Using the moral law as the reality principle renders one guilty of what Kant terms *radical evil*: one does the right thing for the wrong reasons and are thus more in the wrong than those who simply do evil without thinking about it.[9] Kant's condemnation of treating the moral law as the reality principle reveals that he conceives it instead according to the paradigm of the Freudian death drive.

Freud himself does not at all see Kant as his forerunner when he announces the death drive in *Beyond the Pleasure Principle.*[10] In fact, Freud completely misses the significance of the Kantian moral law. In *The Ego and the Id*, he goes so far as to associate it with the logic of the super-ego rather than the death drive (which he had already theorised prior to writing this work). He proclaims, "As a child was once under a compulsion to obey its parents, so the ego submits to the categorical imperative of its super-ego."[11] In a clear reference to the Kantian moral law (identified as the categorical imperative), Freud locates it not as the site of the subject's freedom from external compulsion but in the opposite way, as the index of such compulsion. As Freud sees it here and elsewhere, the moral law is nothing but a manifestation of the superego and thus a vehicle through which the subject impels itself to respect the interests of the society at the expense of its own desire.[12]

Freud's inability to recognise a precursor in the figure of Kant stems ironically from his own refusal to confront the radicality of his own discovery of the death drive a year earlier. Like Kant with the moral law, Freud doesn't quite know what he's onto. He doesn'tt see that the subject's unconscious self-destruction has no erotic force to counterbalance it—there is no opposition of death drive and Eros—or that what he takes as a unifying force does not exist apart from the self-destruction that it allows.

Nonetheless, Freud recognises what Kant does not. The subject's self-destructiveness is the result of its uprooting from animal being, not from a transcendental counter to animality that coexists with animality in subjectivity. In his contrast between the starry heavens above and the moral law within, Kant sets up an ultimately misleading opposition that continues to bedevil those thinking about why the subject does what it does. The problem with this duality is that it enables the subject and the human animal to coexist, which they cannot.

Fortunately for us, the physicists of modernity have not all accepted his separation of the human animal from the self-destructive subject that Kant tacitly accepts. Though they didn't know that they were evincing awareness of the effects of subjectivity, it should not surprise us that the great scientific discoveries of modernity almost always involve scientists taking into account the distortion that the subject introduces into the world that it observes. From Copernicus's counterintuitive thesis that the Earth must revolve around the Sun to Einstein's counterintuitive claim of spatiotemporal relativity to the counterintuitive Many Worlds interpretation (by Niels Bohr and Werner Heisenberg) of the wave function collapse in quantum mechanics, scientists have foregrounded the role that the distortion of subjectivity plays in the act of perception. While they were not explicitly theorising a split subjectivity divided by the cut of signification, they were nonetheless thinking about how their subjectivity intruded in the objective field that they were analysing. They looked back from problems in the data to the perspective from which the data was accumulated. Copernicus's breakthrough occurred when he took the distorting effect of his own movement into account and saw this movement existing in the movement of the sun. What he saw as a purely external movement became the index of the movement of his own subjectivity on earth. Each modern scientific breakthrough occurs in a similar way. Rather than accepting an absolute separation between themselves and external objects, scientists understood that one must work through the role that subjective distortion plays in the visual field in order to gain whatever insight we might find concerning the natural world.

But if the groundbreaking physicists of modernity have taken the disruption of subjectivity to heart, the same cannot be said for modern biology, which is committed to understanding subjectivity as nothing but animality. Modern biology tends to treat subjects as natural genetic organisms instinctually bound to the pursuit of their own genetic good. It is as if physics has worked retroactively from subjectivity to the natural world while biology (at least after Darwin) has worked progressively from the natural world to the subject and thereby disavowed the distortion of subjectivity. From this perspective, the behaviour of the human animal fits comfortably within the natural world that houses it. It is not an excessive figure driven to act against its own good or against the good of the order in which it exists. The point of Darwinian explanations of the subject is to naturalise its excesses, to see the distortion of subjectivity as a result of natural processes rather than as a cause that transcends the limits that natural finitude places on it. But if we look closely at these excesses, their estrangement from natural processes becomes hard to miss.

Extreme Satisfactions

For subjects, both eating and sex become detached from their animal function. Though they continue to fill biological needs, a complete reversal renders biology secondary, so that biology no longer plays a distinct role in the satisfaction that eating and sex provide. To insist on the distinct role that biology plays is to fall into the trap of prioritising the human animal ahead of the self-destructive subject, to believe that subjects can have an unmediated relation to their needs.

The signifier introduces a fundamental distortion in the subject's needs so that it can never approach its needs simply as needs. They become occasions for discovering satisfaction that transcends animal satiation. As Lacan points out in "The Signification of the Phallus," "The phenomenology that emerges from analytic experiencc is certainly of a kind to demonstrate the paradoxical, deviant, erratic, eccentric and even scandalous nature of desire that distinguishes itself from need."[13] The excesses of the desiring subject testify to its fundamental alienation from biological need.

In the case of both eating and sex, the subject's desire often interferes with the satisfaction of biological need. Overeating is the most conspicuous example of this phenomena. When we enjoy the excessive of overeating just for the sake of overeating, we impede the role that eating has in keeping us properly nourished. We eat chocolate doughnut after chocolate doughnut, not in response to biological indications of hunger, but because the act of eating to excess feeds our desire without ever actually satiating it. This process can go on even when we feel absolutely stuffed because the acting of eating in the subject is completely divorced from the need for food. Our desire for food is alienated from our biological need for nourishment.[14] Even if we decide to enjoy a meal of quinoa and spinach, this healthy choice becomes enjoyable not because it best satisfies our biological need but because it testifies to our self-restraint. Of course, we can find pleasure in eating quinoa and spinach for their own sake, but to enjoying eating such a meal requires associating the meal for some type of sacrifice. Once we see the priority of the excessive subject, healthy eating becomes another form of self-denial. Its enjoyment is inextricable from what we give up in order to engage in it.

When one considers sexuality, the divorce between desire and biology becomes even clearer. As women have begun to have children later in life, the idea of sex targeting the moment of ovulation has become popular. Couples calculate when the woman is ovulating, and they have sex at this time in order to maximise the odds of producing a child. If the satisfaction that derives from sex is the satisfaction of reproducing oneself and fulfilling a biological need to reproduce, one would imagine that these targeted sex events would be the most satisfying imaginable. But even if one has not participated in such an activity, it does not require a great feat of imagination to understand that this is not the case. Having sex just to produce a baby actually has the effect of eliminating the enjoyment of the activity. It becomes an onerous duty that one performs, and if there is any enjoyment in it, the enjoyment derives from successfully performing a difficult task—having sex at the precise point of ovulation—not in the act of reproduction. One loses the satisfaction that usually accompanies sex because this satisfaction depends on the divorce between the sex act and reproduction, on the break from all utility and productivity that constitutes the nonreproductive sex act. Purely reproductive sex is akin to having a diet consisting solely of kale.

If we define perversion as the derailing of an instinct from its purely biological function, then all subjects are perverse in their eating and their sexuality.[15] The satisfaction that both eating and sex produce for the speaking subject stems from the distance between these acts and the biological need that they fulfil. Those who enjoy overeating binge not on the foods that will best ensure their survival but on those that will potentially kill them. They eat, for example, cookies, cake, and ice cream. They seek out sugars and simple carbohydrates. Those who devote themselves to sex do not confine themselves to acts that will best ensure reproduction but rather explore ones that evince the distance between sex and reproduction. They try masturbation, fellatio, cunnilingus, anal sex, group sex, and multiple variations. Just as no overeater has ever binged on spinach, no sex addict has ever overdosed on reproductive sex.

Attempts to explain overeating or sex addiction with recourse to biological causes are commonplace. For instance, rather than theorising obesity as the result of the subject's excessiveness, evolutionary theorists see it as a natural development of an evolutionary mismatch. We developed proclivities for a situation that no longer existed, and the evolved characteristics that once worked to our advantage now threaten our survival. According to evolutionary biologist Daniel Lieberman, "Millions of years of evolution favored ancestors who craved energy-rich foods, including simple carbohydrates like sugar which used to be rare, and who efficiently stored excess calories as fat."[16] Evolution prepared the human animal properly, but cultural changes worked faster than natural selection, resulting in widespread obesity in contemporary society. Scholars in all fields and most of the general public tend to accept these interpretations of the subject in terms of the animal as definitive.

Explanations for the subject's sexual excesses tend to follow those of its culinary excesses. Males cheat on their spouses because their genes want to spread themselves as widely as possible, whereas females opt for monogamy because their genes see this as the best strategy for helping their offspring to survive. In *How the Mind Works*, Steven Pinker offers a standard explanation for male promiscuity in terms of adaptation. He writes, "The desire for sexual variety is an unusual adaptation, for it is insatiable.... The more women a man has sex with, the more offspring he leaves; too much is never enough. That gives men a limitless appetite for casual sex partners (and perhaps for the commodities that in ancestral environments would have led to multiple partners, such as power and wealth)."[17] Natural selection can explain even the most extreme sexual acts because participation in them testifies to the fitness of the participant. There is no excess so excessive that it cannot seem to be the result of the mechanism of adaptation.

As if in an effort to validate this dictum, evolutionary biologists have spent considerable energy trying to discover the adaptive reasons for rape. Though many colleagues attacked them for their research, Randy Thornhill and Craig Palmer studied the question of the causes for the evolution of rape. The result of their research, *A Natural History of Rape: Biological Bases of Sexual Coercion*, leaves open two different possibilities—one preferred by each of the coauthors.[18] For Thornhill, rape is a result of adaptation, whereas for Palmer, it is a byproduct of the adaptive process. Though Thornhill and Palmer take pains to point out that the goal of their research is to arrive at better mechanisms for the prevention of rape, the very concept of rape as a biological inheritance landed them immediately in political hot water.

Nonetheless, we should applaud Thornhill and Palmer for taking the mechanism of evolutionary explanation to its endpoint. If it is the case that all of the subject's excesses are simply developments of its animality, this logic must apply to rape as well. Even if the pursuit of the evolutionary solution to rape seems politically inadvisable, the emergence of this line of thought follows directly from the fundamental assumption of the consideration of humans as animals rather than as subjects. If subjects are nothing but animals, all their actions, even the most excessive, must be natural and thus fit within the adaptive framework. If Thornhill and Palmer did not theorise rape as an evolutionary adaptation, someone else would have done so. From this perspective, no excess is so extreme that it departs from animality, and this gives rise to the demand for a natural explanation. What this explanation spares the explainers is the necessity of seeing the complicity of not just the rapist but of all subjects in the society that produces rape as a possible practice.

It is certainly possible to conceive of all the excesses associated with subjectivity as developments or offshoots of natural selection occurring in the human animal. Even when the case seems most strained (as in the case of rape), one can nonetheless put forth a plausible argument that wins adherents. But these arguments ultimately seem inadequate to the phenomena they attempt to account for because they operate under the assumption that subjects, qua human animals, act for the sake of a fundamental self-interest. According to this Neo-Darwinist perspective, they seek to propagate themselves—or more specifically, their own genetic material—and this instinct for self-preservation and perpetuation trumps all other factors in the activity of the subject.

The excesses of subjectivity that strain the explanation of self-reproduction are not just exceptional cases found in a few isolated subjects who haven't managed to get with the program thanks to some miscoding in their genes. Instead, these excesses appear in every subject and, what's more, they provide the means through which subjects satisfy themselves. The excesses of subjectivity are not akin to unnecessary organs like the appendix but rather give the subject its reason for continuing to exist. Whether it is a sexual practice, a type of food, or a particular over-valued object, the excessive attachment to what doesn't serve its self-interest satisfies the subject in a way that self-reproduction does not. The irony of subjectivity is that if subjects could only reproduce themselves without also destroying themselves they would have no desire to do so. They continue to exist as a result of the satisfaction that they derive from how they perversely violate the imperative of self-reproduction. Their self-reproduction depends on their refusal to follow any genetic impulse to reproduce themselves.

If subjects could simply pursue their own interest or their own good (or the good of their genes), they would belong to the world in which they exist. But the subject fails to belong to the natural world because it constantly deviates from what would serve its good or the good of its world. The subject's failure to belong to the natural world derives from its inherent self-destructiveness. One must view this self-destructiveness, in spite of the horrors it has produced, as the key to the subject's singularity and thus the sine qua non of subjectivity.

Even though the subject's excess is responsible for countless horrors in the world, our response to this excess must involve its embrace. Without the excess that denaturalises the human animal, the subject would lack the singularity that gives its existence a value. When we misconceive the subject's self-destruction as a natural phenomenon and try to explain it in those terms, we imagine that we might normalise this excess. But the subject's self-destructiveness is both the problem and the solution. We must revaluate this self-destruction as the basis for the subject's ethical being, as Kant already indicates. The path towards doing so leads through Franz Kafka.

From Anorexia to Immediacy

As Gilles Deleuze and Félix Guattari point out, Franz Kafka often shows the barrier between the animal and the human breaking down.[19] For them, the paradigmatic Kafka story is *The Metamorphosis*. In this story, Gregor Samsa literally embodies our ability to become-animal and thereby eviscerates the hard and fast distinction. But Kafka's greatest insight into the role that animality plays for the subject occurs in a story where he highlights the distinction between the human and the animal rather than eliminating it. This is what occurs in "The Hunger Artist." Though it portrays a speaking subject on display in a cage like an animal at a zoo, the story does not spend much time discussing animality, but the conclusion of "The Hunger Artist" brings out this distinction while at the same time revealing the seductiveness of the fantasy of animal immediacy.

Kafka's story recounts the fame and subsequent demise of a hunger artist, someone who starves himself to entertain spectators. While the hunger artist and others of his profession are the height of their popularity, they attract hundreds of spectators and incite a great deal of enthusiasm. Kafka presents the hunger artist as satisfied in the art of not eating, and the enjoyment that the spectators find seems just to supplement this inherent satisfaction of the artist. It is a satisfaction that comes from defying the natural impulse to eat rather than submitting to it. The hunger artist and the spectators foreground the unnatural excess of the subject, even if this excess manifests itself through abstention.

At first, it seems strange that starving oneself might be considered an art rather than a disorder, but in fact all arts require some form of extreme self-sacrifice. Although the romantic image of the tortured artist is largely mythological, it is nonetheless the case that every work of relies on spending a great deal of time both training to be an artist and performing one's art. Usually, this effort is lifelong. The artist must at the very least sacrifice a large portion of her or his life for the sake of the art, but usually the sacrifice goes beyond giving up food. In the most extreme case, the artist, such as Vincent Van Gogh or Sylvia Plath, goes mad in pursuit of the great work of art. Even those artists who live stable and successful lives, such as Ousmane Sembène or Toni Morrison, nonetheless sacrifice the lion's share of their lifetimes for the sake of producing their art. Of course, the audience for the artwork doesn't see this sacrifice directly, but it is visible in the suffering that the artwork reveals. The artist's sacrifice produces a sublimity in the art—a moment in which the work of art transcends the concerns of everyday life—and it is what spectators want to experience when they engage with the work of art. But at the same time, the excessiveness of the artwork traumatizes the spectator who sees it. Looking at the great work of art confronts the subject with the excessiveness of its own subjectivity, which is why regarding a great work of art is always traumatic. What draws the spectator in at the same time repulses the spectator.

The trauma of spectatorship is especially apparent in "The Hunger Artist." One sees a superhuman feat—someone not eating for weeks—and yet there is something horrific about the emaciated body of the hunger artist. Despite the crowds that flock to watch the hunger artist's "skeleton thinness," as Kafka puts it, "many people had regretfully to keep away from his exhibitions, because the sight of him was too much for them."[20] It is enjoyable to watch the excessive achievement of the hunger artist, his courageous defiance of his natural inclinations, but this is an enjoyment rooted in trauma because its divorce from nature is self-evident.

This is why, as the narrator of the story points out, that officials place a mandatory limit of forty days on the fast. No hunger artist can continue past the forty day mark and attempt to establish an individual record, and the hunger artist feels this limit as a real injustice. But officials impose it in order to contain the trauma (and the enjoyment) that the art of starvation unleashes. But eventually the effort at containment becomes unnecessary: the spectators find another form of enjoyment that spares them the unnatural excess of the hunger artist.

The end of Kafka's story finds the hunger artist as a forgotten sideshow attraction at a circus, and when the proprietor finally puts an end to his act, there are no more spectators. They all flock to see the panther, who replaces the hunger artist in his cage. The panther also horrifies the spectators but not so that they look away. Instead, in the vision of the spectators, the animal embodies a perfectly immediate satisfaction. In the final paragraph of the story, Kafka writes,

> The panther was all right. The food he liked was brought to him without hesitation by the attendants; he seemed not even to miss his freedom; his noble body, furnished almost to the bursting point with all that it needed, seemed to carry freedom around with it too; somewhere in its jaws it seemed to lurk; and the joy of life streamed with such ardent passion from his throat that for the onlookers it was not easy to stand the shock of it. But

> they braced themselves, crowded around the cage, and did not want ever to move away.[21]

The panther provides a similar thrill to that of the hunger artist, though the distinction that Kafka describes captures what's at stake in the fantasy of animal immediacy as a vehicle for the subject's excess. Unlike the hunger artist, the panther, because it is not a desiring subject, seems to have the ability to have its needs fulfilled without destroying itself in the process. The panther satisfies itself by eating to sustain itself, whereas the subject satisfies itself by not eating to destroy itself. But this contrast only exists through the image that subjectivity itself produces.

While the traumatic excess of the hunger artist forces some to turn away, the exciting shock of the panther, though it disturbs spectators, keeps them glued to its image. The fantasy of animal immediacy enthrals the onlookers in a way that the excessive abstinence of the hunger artist does not. The panther enables the spectator to associate its excessiveness with its nature, and in this way, the traumatic distortion that mediation produces in the subject disappears. As one watches the panther, one sees the fantasy of one's own former natural being.

The stunning achievement of Kafka's story is that it captures precisely why we prefer to think of the excess of subjectivity in animal terms. Animality enables us to imagine that our desires are really just needs that we can satisfy if we find the appropriate objects. Animality offers an image of wholeness that obscures the hole in the world that our subjectivity produces. We are drawn to animal terms to describe subjectivity for the same reason that the crowds abandon the hunger artist and flock to the panther at the end of "The Hunger Artist." But the seeming immediacy of the animal will always leave us dissatisfied, despite the sense of wholeness that it communicates. Even if Kafka portrays the abandonment of the hunger artist by those who flock to see the panther, the existence of the story itself indicates a desire to move beyond the perfect image of the panther.

The problem is not just that the speaking subject cannot return to its former animal form but rather that this animal form takes on the veneer of natural immediacy only in retrospect, only after one has entered into mediation. The more that we break down the barrier between subject and animal, the more we lose sight of the unnatural form of our way of existing. The naturalness of the relation that exists prior to mediation is itself a product of our unnatural subjectivity. The natural world is the product of subjectivity as much as subjectivity is a product of the natural world. To talk about the human animal or about the natural world as a separate and distinct world is to miss the subjectivity that enables us to see this separation. There is no return to immediacy because there is no immediacy to return to. But the fantasy of a lost immediacy has the capacity to blind us to the privilege that self-destruction plays in our existence.

When Kant came to the end of the *Critical of Practical Reason*, he wanted to offer readers a sense of the magnitude of the moral law and its impact on the subject. But in order to do so, he had to begin by imagining himself as an insignificant natural being confronting the starry heavens above. With this gesture, Kant blinds himself to the excessiveness of his own discovery: it is only through the transcendence that the moral law enacts that he could imagine his own insignificance. His look to the heavens was an attempt to avoid noticing just how excessive the subject of the moral law was. We imagine ourselves as animals—even in part—in order to tame the excess of subjectivity. It is only by recognising that our animality follows from our excessiveness that we can confront the ramifications of our own disruption of the world.

Notes

1. Immanuel Kant, *Critique of Practical Reason*, in *Practical Philosophy*, trans. and ed. Mary J. Gregor (New York: Cambridge University Press, 1996), 269; Kant's emphasis.
2. Kant's assertion of the primacy of finitude contrasts with Georg Wilhelm Friedrich Hegel, who sees the subject's infinitude as constitutive of its ability to recognize that it is finite. But this assertion earns Kant the approbation of Martin Heidegger, who finds in him a more proximate thinker than Hegel, precisely because Kant insists on a finitude of subjectivity that it cannot transcend. The moral law lifts the subject, for Kant, but it doesn't allow the subject to escape its primary finitude.

3. Kant contrasts obedience of the moral law, which is the source of freedom, with all other sorts of obedience. He indicates this difference with the terms *ideal* and *idol*. In the *Metaphysics of Morals*, he states, "Kneeling down or prostrating oneself on the ground, even to show your veneration for heavenly objects, is contrary to the dignity of humanity, as is invoking them in actual images; for you then humble yourself, not before an *ideal* presented to you by your own reason, but before an *idol* of your own making." Immanuel Kant, *The Metaphysics of Morals*, in *Practical Philosophy*, trans. and ed. Mary J. Gregor (Cambridge: Cambridge University Press, 1996), 559; Kant's emphasis. Adherence to the moral ideal frees the subject from the lure of all idols, including those given by other members of one's society, which otherwise govern the subject's existence.
4. This is, in effect, the critique of Kant that object oriented ontology puts forward. For those that take up this position, Kant's insistence on the priority of the subject represents the major wrong turn in the history of modern Western philosophy. Once one takes this turn, one loses touch with the real object world and immerses oneself in the trap of representations.
5. The priority that Kant gives to animal being testifies to an investment in the primacy of what Jacques Lacan calls the imaginary order. By understanding the relationship between the subject and the world in terms of the imaginary, one misses the distorting effect of the signifier's mediation. As Lacan puts it in his *Seminar IV*, "The imaginary relation, which is an essentially alienated, interrupted, slowed, inhibited, and most often inverted relation, profoundly misrecognizes the relationship of speech between the subject and the Other, the big Other." Jacques Lacan, *Le Séminaire, livre IV: La relation d'object, 1956-1957*, ed. Jacques-Alain Miller (Paris: Seuil, 1994), 12. Taking the imaginary relation as a starting point for his theorizing forces Kant to miss the impossibility of the speaking subject simply acting like an animal.
6. This recognition that our infinitude—our capacity to think beyond the finite givenness of our existence—is what enables our finitude to become a problem for us is an insight that appears in Hegel's reworking of the Kantian philosophy.
7. Although Freud distinguishes between Eros and death drive, Lacan insists that "every drive is virtually a death drive." Jacques Lacan, "The Position of the Unconscious," in *Écrits: The First Complete Edition in English*, trans. Bruce Fink (New York: W. W. Norton, 2006), 719.

8. The moral law is the basis for freedom because it opens up a space between what I'm inclined to do and what I do. As Alenka Zupančič points out in her panegyric to Kantian morality, "The defining feature of a free act … is precisely that it is entirely foreign to the subject's inclinations." Alenka Zupančič, *Ethics of the Real: Kant, Lacan* (New York: Verso, 2000), 23. Without the uprooting encounter with the moral law, the subject remains just what it is and follows the inclinations given to it.
9. See Joan Copjec, ed. *Radical Evil* (London: Verso, 1996).
10. See Sigmund Freud, *Beyond the Pleasure Principle*, trans. James Strachey, in *The Standard Edition of the Complete Psychological Works of Sigmund Freud*, ed. James Strachey (London: Hogarth Press, 1955), vol. 18, 1-64.
11. Sigmund Freud, *The Ego and the Id*, trans. James Strachey, in *The Standard Edition of the Complete Psychological Works of Sigmund Freud*, ed. James Strachey (London: Hogarth, 1961), vol. 19, 48.
12. In contrast to Freud, Lacan does have a sense of the radicality of Kant's insight in the discovery of the moral law. When he writes "Kant with Sade," Lacan grasps that Kant performs the revolutionary step of turning morality away from the good, which is where it had been lodged since Aristotle. As Lacan sees it, even the pleasures that moral acts bring for the subject has the effect of disqualifying them as moral acts. Lacan states that rather than elevating these pleasures the reference to the good "renders these pleasures less respectable." Jacques Lacan, "Kant with Sade," in *Écrits: The First Complete Edition in English*, trans. Bruce Fink (New York: W. W. Norton, 2006), 647.
13. Jacques Lacan, "The Signification of the Phallus," in *Écrits: The First Complete Edition in English*, trans. Bruce Fink (New York: W. W. Norton, 2006), 579.
14. Those who eat healthily and claim to enjoy this food seem to confound this analysis. But no one can simply eat healthily for biological reasons. By doing so, one enjoys a different form of self-destruction than the eater of doughnuts. The enjoyment of the health-food junkie derives from the sacrifice of the pleasure in eating tasty foods. With every spear of asparagus or crown of broccoli, one takes satisfaction from the ice cream and Twinkies that one is not eating. There is no path to a purely nutritive eating process.
15. This is how Freud defines sexual perversion, though he insists at the same time that we cannot use it as a critique, precisely because it is nearly ubiquitous in human sexuality. He writes, "Perversions are sexual activities which either (*a*) extend, in an anatomical sense, beyond the regions

of the body that are designed for sexual union, or (*b*) linger over the immediate relatons to the sexual object which should normally be traversed rapidly on the path towards the final sexual aim." Sigmund Freud, *Three Essays on the Three of Sexuality*, trans. James Strachey, in *The Standard Edition of the Complete Psychological Works of Sigmund Freud*, ed. James Strachey (London: Hogarth Press, 1953), vol. 7, 149-50.

16. Daniel Lieberman, *The Story of the Human Body: Evolution, Heath, and Disease* (New York" Pantheon, 2013), 16.
17. Steven Pinker, *How the Mind Works* (New York: W. W. Norton, 1997), 473. Pinker claims that this adaptation for promiscuity is confined completely to men and that women have little genetic interest at all in multiple partners. But fellow evolutionary psychologist David M. Buss counters by pointing out the adaptive benefits of affairs for women. He states, "The economics of the mating market dictate that women can secure genes from an affair partner that are superior to those of her regular partner, at least in principle. A highly desirable man is often willing to have a brief encounter with a less desirable woman, as long as she does not burden him with entangling commitments. Indeed, if there were no costs, at optimal female mating strategy, in the ruthless currency of reproductive fitness, would be to secure reliable investments from her husband and superior genes from an affair partner." David M. Buss, *The Evolution of Desire: Strategies of Human Mating*, rev. ed. (New York: Basic Books, 2003), 235-36. The problem with each of these arguments lies in the fact that the only way that one can decide between them is on the basis of what one considers their political desirability. One might even say that these opposed arguments reveal the antinomies of evolutionary reason. For an unparalleled critique of the sexism informing evolutionary psychology, see Mari Ruti, *The Age of Scientific Sexism: How Evolutionary Psychology Promotes Gender Profiling and Fans the Battle of the Sexes* (New York: Bloomsbury, 2015).
18. Randy Thornhill and Craig T. Palmer, *A Natural History of Rape: Biological Bases of Sexual Coercion* (Cambridge, MA: MIT Press, 2000).
19. See Gilles Deleuze and Félix Guattari, *Kafka: Toward a Minor Literature*, trans. Dana Polan (Minneapolis: University of Minnesota Press, 1986).
20. Franz Kafka, "The Hunger Artist," trans. Willa Muir and Edwin Muir, in *The Complete Stories*, ed. Nahum M. Glatzer (New York: Schocken, 1971), 270.
21. Kafka, "Hunger Artist," 277.

Bibliography

Buss, David M. 2003. *The Evolution of Desire: Strategies of Human Mating.* Revised ed. New York: Basic Books.

Copjec, Joan, ed. 1996. *Radical Evil.* London: Verso.

Deleuze, Gilles, and Félix Guattari. *Kafka: Toward a Minor Literature.* Translated by Dana Polan. Minneapolis: University of Minnesota Press, 1986.

Freud, Sigmund. 1953. *Three Essays on the Three of Sexuality*, Trans. James Strachey. In *The Standard Edition of the Complete Psychological Works of Sigmund Freud*, ed. James Strachey, vol. 7, 123-245. London: Hogarth Press.

———. 1955a. *Beyond the Pleasure Principle*, Trans. James Strachey. In *The Standard Edition of the Complete Psychological Works of Sigmund Freud*, ed. James Strachey, vol. 18, 1-64. London: Hogarth Press.

———. 1955b. *The Ego and the Id*, Trans. James Strachey. In *The Standard Edition of the Complete Psychological Works of Sigmund Freud*, ed. James Strachey, vol. 19, 1-66. London: Hogarth.

Kafka, Franz. 1971. The Hunger Artist, Trans. Willa Muir and Edwin Muir. In *The Complete Stories*, ed. Nahum M. Glatzer, 268-77. New York: Schocken.

Kant, Immanuel. 1996a. *Critique of Practical Reason.* In *Practical Philosophy*, Trans. and ed. Mary J. Gregor. New York: Cambridge University Press.

———. 1996b. *The Metaphysics of Morals.* In *Practical Philosophy*, Trans. and ed. Mary J. Gregor. Cambridge: Cambridge University Press.

Lacan, Jacques. 1994. In *Le Séminaire, livre IV: La relation d'object, 1956-1957*, ed. Jacques-Alain Miller. Paris: Seuil.

———. 2006. *Écrits: The First Complete Edition in English.* Trans. Bruce Fink. New York: W. W. Norton.

Lieberman, Daniel. 2013. *The Story of the Human Body: Evolution, Heath, and Disease.* New York: Pantheon.

Pinker, Steven. 1997. *How the Mind Works.* New York: W. W. Norton.

Ruti, Mari. *The Age of Scientific Sexism: How Evolutionary Psychology Promotes Gender Profiling and Fans the Battle of the Sexes.* New York: Bloomsbury, 2015.

Thornhill, Randy, and Craig T. Palmer. 2000. *A Natural History of Rape: Biological Bases of Sexual Coercion.* Cambridge, MA: MIT Press.

Zupančič, Alenka. 2000. *Ethics of the Real: Kant, Lacan.* New York: Verso.

16

Afterward: COVID-19 and the Environment

Clint Burnham and Paul Kingsbury

The editors' final preparation of this volume took place in the spring and summer of 2020—during, that is, the unprecedented global pandemic of the novel coronavirus SARS-CoV-2 and acute respiratory disease COVID-19. As the world gradually shut down, as schools, factories, and restaurants, first in East Asia, then Europe, then North America, Africa, South Asia, Oceania, and South America, all closed or dramatically pivoted to online operations, an early ray of light, of perhaps desperately sought for hope, was the temporary abatement of pollution due to industry, commuting, and tourist travel. The Himalayas were suddenly visible from Punjab rooftops,[1] NASA reported that pollution in China had plummeted,[2] and deer were spotted wandering down city streets.[3]

C. Burnham (✉)
English Department, Simon Fraser University, Vancouver, BC, Canada
e-mail: clint_burnham@sfu.ca

P. Kingsbury
Faculty of Environment, Simon Fraser University, Burnaby, BC, Canada
e-mail: kingsbury@sfu.ca

C. Burnham, P. Kingsbury (eds.), *Lacan and the Environment*, The Palgrave Lacan Series, https://doi.org/10.1007/978-3-030-67205-8_16

Furthermore, as governments responded to COVID-19 with huge outlays of social programs and quickly moved to support industries and communities in need, activists pointed out that such state actions demonstrated the resurgent possibilities for Keynesian public outlays.[4]

And yet, as Alessandra Capperdoni points out in a blog post, such good news may be less so: "The pollution of skies and seas has suddenly decreased. The canals of Venice are transparent again. Humans cannot walk freely in the cities but can observe the silent spectacle of artistic beauty from their homes."[5] That is, the Himalayas or Venetian canals merely provide more online content for our screens, for the digital sublime. And, even as governments have freed up money for social programs, amidst an international search for a vaccine, the denialism around the virus, from populist leaders like Brazil's Jair Bolsonaro and the U.S.'s Donald Trump to popular anti-maskers suggest connections with pre-existing climate skeptics and anti-vaxxers. Todd McGowan, however, suggests that disasters like COVID-19 "show us the universality that bond us together."[6]

What, then, might be the Lacanian lesson of COVID-19 and the environment? Perhaps we can think of this in terms of Lacan's argument for the "not-all," or a form of universalism that derives from the non-relation (argued in his *Encore* seminar and articulated in the formulas of sexuation). That is, for both climate change and COVID-19, there exists a liberal argument, often made by well-meaning scientists and activists, that "we are all in this together." We are all on this planet, we are all in danger of catching the virus. Even Tom Hanks. Therefore, we should take dramatic action to reduce our carbon production such as eating less meat and flying less. Therefore, we should follow social distance guidelines, wash our hands, and wear masks. In the Lacanian matrix, this proposition is signaled with the matheme $\forall x\ \Phi x$—that is, all, or $\forall$, are subject to castration, Φ. All lack: in these cases, the susceptibility to climate change or COVID-19.

But in this logic, there is always, necessarily, an exception to the all—the all qua totality must generate an exception, signified in the matheme $\exists x\ \Phi x$, or "there exists someone who is not subject to that lack." This logic of exception, which is informed by Freud's figure of the primal father in *Totem and Taboo,* is also the logic of a subject who abjures the collectivity. It is a logic that coordinates social inequities that doom some or many to live on an overheated planet and to suffer from inadequate health care while others can buy Teslas, and, like the celebrities singing "Imagine," own and perhaps even sometimes dwell in spacious mansions. The logic of the exception is a masculine logic. Here, the newly found distanciated admiration of pristine, or, perhaps more accurately, pre-castrated mountains and canals recalls Alenka Zupančič's (2017:79) alignment of masculine sexuation with a natural environment that "keeps standing there not as an impenetrable Real in itself, but as the Imaginary, which we can see, like, and love, but which is, at the same time, somewhat irrelevant."

On the right side of Lacan's formula of sexuation, next to the logic of exception, we find the feminine matheme $\overline{\exists x}\ \overline{\Phi x}$, or "there does not exist anyone who is not subject to castration." According to this logic, Lacan argues, we arrive at the logic of the not-all, with the matheme $\overline{\forall x}$ Φx, that is, not-all are subject to castration or lack. Such a logic points towards an immanently incomplete and out-of-joint subject that lacks a determinate relation to an external limit or outside. Such a subject also recalls a figure who is able to radically disturb pre-existing states of regulated wholeness, equilibrium, and completeness. A potential figure here is Gaia wherein the Earth is understood as a living entity or biosphere. According to Lucas Pohl (2020), against the usual readings of Gaia as exemplifying a "harmonious figure of whole-ism", Gaia can actually enable us to traverse the masculine fantasy of mOther Nature and "any notion of Oneness" by "her ability to pave the way towards the void taking place at the heart of reality." For Pohl, crucial here are Gaia's objects of ruination (e.g. moss growing like carpet in Detroit's abandoned buildings) that reveal the impossibility of conceptualizing nature and culture as binary opposites.

To be sure, viruses are *un*dead insofar as they cannot self-replicate (they need a host and its cellular infrastructure to do this) and are more akin to robots than living organisms. Finally, COVID-19 has ushered in new forms of uneasiness in culture, especially in the corridors of consumptions such as accidently walking the wrong way down one-way grocery store aisles or being reminded that one is trespassing the 2 meter/6 feet social distancing rule in a store by a subtle gesture (a quiet cough of politeness in one case). Finally, COVID-19 has ushered in new forms of ruination in the built environment such as sun-bleached kid-drawn posters of support for frontline workers in street-facing windows, or the soiled, decaying, and not to be touched by bare hands masks on floors, as well as smudged, scuffed, ripped stickers and masking tape guiding shoppers where to wait on sidewalks.

A last comment on these positions can perhaps help here. The logic of exception is that of the fantasy we have been inundated with during COVID-19, that now with fewer airline flights and less commuter traffic (apart from bikes), now we are seeing what the world can be in a greener future. *But this fantasy is precisely that of what Žižek has called the impossible gaze.* The idea of viewing our own birth, or death: this is the fantasy that subtends both deep ecology (humans are the real virus—the Anthropocene) and such literary-visual productions as *The World Without Us*. If the fantasy of a greener, COVID-19 planet adheres to this masculine logic, this logic of the exception, what of the logic of the not-all, the feminine? Here we can think of nature as COVID-19. Consider the virus's uncanny ability to survive by interlacing itself with cultural objects such as handrails, reusable grocery bags, and playgrounds. That is, COVID-19 foregrounds the *un*naturalness of nature, akin to "the *un*canny or *un*conscious as something inherently distorted through the seal of parasitism" (Pohl, 2020), with its Lego-like or car assembly plant-like cycles of infection involving attachment and entry, membrane fusion and release, translation, transcription, replication, trafficking, assembly, and eventual release of mature virions. Perhaps we *can* learn to love this virus that is "in us more than us": remember when you first saw the colourful image of the COVID-19 virus, with the red, orange, and yellow spikes that, crown-like, surrounded the grey membrane? How beautiful, or cute, like a dog's chew toy, it looked. If few can love COVID-19, then do we deserve to love the environment?

Notes

1. https://www.abc.net.au/news/2020-04-09/himalayas-visible-indiapollution/12136856
2. https://earthobservatory.nasa.gov/images/146362/airborne-nitrogen-dioxide-plummets-over-china

3. https://www.theguardian.com/world/2020/mar/22/animals-cities-coronavirus-lockdowns-deer-raccoons
4. https://ecojustice.ca/canadas-covid-19-response-a-blueprint-for-climate-action/
5. http://www.lacansalon.com/listening-to-covid-19/viral-time/the-voracious-desire-of-covid-19
6. 6. Todd McGowan, "Why the Right Hates Masks", https://www.youtube.com/watch?v=vo9ik8nGEmY July 26, 2020.

Bibliography

Pohl, Lucas. 2020. Ruins of Gaia: Towards a Feminine Ontology of the Anthropocene. *Theory, Culture & Society* 37 (6): 67–86.

Zupančič, Alenka. 2017. *What Is Sex?* Cambridge, MA: MIT Press.

Index[1]

[1] Note: Page numbers followed by 'n' refer to notes.

C. Burnham, P. Kingsbury (eds.), *Lacan and the Environment*, The Palgrave Lacan Series, https://doi.org/10.1007/978-3-030-67205-8

www.ingramcontent.com/pod-product-compliance
Lightning Source LLC
Chambersburg PA
CBHW050512261225
37173CB00059BA/110

* 9 7 8 3 0 3 0 6 7 2 0 4 1 *